Liaison Functions

Core Programming Topic at the 2012 AIChE Annual Meeting

Pittsburgh, Pennsylvania, USA
28 October - 2 November 2012

ISBN: 978-1-62276-731-1

Printed from e-media with permission by:

Curran Associates, Inc.
57 Morehouse Lane
Red Hook, NY 12571

Some format issues inherent in the e-media version may also appear in this print version.

Printed by Curran Associates, Inc. (2013)

For permission requests, please contact AIChE
at the address below.

AIChE
3 Park Avenue
New York, NY 10016-5991

Phone: (203) 702-7660
Fax: (203) 775-5177

www.aiche.org

Additional copies of this publication are available from:

Curran Associates, Inc.
57 Morehouse Lane
Red Hook, NY 12571 USA
Phone: 845-758-0400
Fax: 845-758-2634
Email: curran@proceedings.com
Web: www.proceedings.com

TABLE OF CONTENTS

(5a) Introduction to the MBA

Ashlie Prioleau, Carnegie Mellon University, New York, NY

This is a tutorial-like presentation of the MBA In A Day session. An invited speaker will give an overview of the title subject, who, tentatively, is faculty member from a regional business school.

The session will be of interest to those who want to learn about MBA programs and content: Those who may be considering enrolling in an MBA program, those who want to understand the type of information that is covered in an MBA program, and those who would benefit from having an overview of the content of a typical MBA program, for example. The speakers are subject matter experts who will take into account the interests and background of chemical engineers in the way that they present the information. The average student who is enrolled in an MBA program has six years of experience and is 29 years old. Students with engineering or science backgrounds comprise 35 to 40% of MBA programs. As a consequence, the session will be highly of interest to young professionals.

Ashlie Prioleau graduated from Carnegie Mellon University in 2011 with a Bachelor's degree in Chemistry, a concentration in Inorganic Chemistry, and a minor in Hispanic Studies. At Carnegie Mellon, Ashlie conducted inorganic research on peptide nucleic acids and alkali metals. She was awarded a Howard Hughes Medical Institute (HHMI) grant and honored as an HHMI scholar for her research.. After working in a number of administrative positions in student affairs and higher education, Ashlie decided to explore non-scientific opportunities. Currently, she is pursuing a Masters degree in Higher Education Management at the University of Pittsburgh while working in the Masters Admissions Office at the Tepper School of Business. In this position, she advises students with technical backgrounds on how to apply for a Masters in Business Administration and/or Computational Finance.

(5b) Entrepreneurship and Innovation

Philip G. Boyer, Engineering Entrepreneurship Minor, The Pennsylvania State University, University Park, PA and Robert Beaury, Engineering Entrepreneurship, The Pennsylvania State University, University Park, PA

This is a tutorial-like presentation of the MBA In A Day session.

Philip G. Boyer, M.B.A. - Director of the Engineering Entrepreneurship Minor, The Pennsylvania State University. Professor Boyer currently teaches entrepreneurship courses and directs the E-Ship Minor at University Park. Prior to Penn State, Boyer was a marketing product brands manager with the Hershey Company and has since been involved with several start-up companies as a director/owner. Additionally, Phil has successfully built and sold an upscale golf course. Phil's primary area of expertise is in marketing and strategic business planning.

Robert Beaury (Bob) is currently an instructor of Entrepreneurship and Leadership at Penn State University's College of Engineering. His classes focus on teaching engineers and business students how to create and market innovative products. Bob served as the CEO of Broadband Networks, Inc. (BNI) for ten years and grew BNI to be one of the leading suppliers of distance learning networks. During his year career, Bob has been directly involved in the creation, development and market introduction of more than 100 new products. His expertise includes building start-up companies and the practical implementation of innovation and product development techniques for cost-conscious, results oriented, organizations.

The session will be of interest to those who want to learn about MBA programs and content: Those who may be considering enrolling in an MBA program, those who want to understand the type of information that is covered in an MBA program, and those who would benefit from having an overview of the content of a typical MBA program, for example. The speakers are subject matter experts who will take into account the interests and background of chemical engineers in the way that they present the information. The average student who is enrolled in an MBA program has six years of experience and is 29 years old. Students with engineering or science backgrounds comprise 35 to 40% of MBA programs. As a consequence, the session will be highly of interest to young professionals

(5c) Accounting, Finance, and Statistics I

Makarand (Raj) Joshi, Organizational Strategy Management and Operations, Merck & Co., Inc., Whitehouse Station, NJ

This is a tutorial-like presentation of the MBA in a Day session. Please attend both talks, Accounting, Finance, and Statistics I and II, if you want to learn about all three subjects, which will be divided across the two talks. All three subjects will not be covered in each of the two talks.

The speaker is Makarand (Raj) Joshi, who is currently Lean Six Sigma Lead for Global Science, Technology, and Commercialization organization at Merck & Co. in New Jersey. He has extensive experience in R&D and Manufacture of Pharmaceuticals and Medical Devices and has led entire gamut of activities from Concept Identification through Product Development to Launch and Supply. Raj has a PhD in Chemical Engineering from Cornell University and started his career in research. Subsequently he moved to development and then to manufacturing. Mid way in his career he obtained an MBA degree from Lehigh University. He is a certified Black Belt and is currently pursuing a Master Black Belt certification.

The session will be of interest to those who want to learn about MBA programs and content: Those who may be considering enrolling in an MBA program, those who want to understand the type of information that is covered in an MBA program, and those who would benefit from having an overview of the content of a typical MBA program, for example. The speakers are subject matter experts who will take into account the interests and background of chemical engineers in the way that they present the information. The average student who is enrolled in an MBA program has six years of experience and is 29 years old. Students with engineering or science backgrounds comprise 35 to 40% of MBA programs. As a consequence, the session will be highly of interest to young professionals.

(5d) Marketing

Robert Beaury, Engineering Entrepreneurship, The Pennsylvania State University, University Park, PA and Philip G. Boyer, Engineering Entrepreneurship Minor, The Pennsylvania State University, University Park, PA

This is a tutorial-like presentation of the MBA In A Day session.

Robert Beaury (Bob) is currently an instructor of Entrepreneurship and Leadership at Penn State University's College of Engineering. His classes focus on teaching engineers and business students how to create and market innovative products. Bob served as the CEO of Broadband Networks, Inc. (BNI) for ten years and grew BNI to be one of the leading suppliers of distance learning networks. During his year career, Bob has been directly involved in the creation, development and market introduction of more than 100 new products. His expertise includes building start-up companies and the practical implementation of innovation and product development techniques for cost-conscious, results oriented, organizations.

Philip G. Boyer, M.B.A. - Director of the Engineering Entrepreneurship Minor, The Pennsylvania State University. Professor Boyer currently teaches entrepreneurship courses and directs the E-Ship Minor at University Park. Prior to Penn State, Boyer was a marketing product brands manager with the Hershey Company and has since been involved with several start-up companies as a director/owner. Additionally, Phil has successfully built and sold an upscale golf course. Phil's primary area of expertise is in marketing and strategic business planning.

The session will be of interest to those who want to learn about MBA programs and content: Those who may be considering enrolling in an MBA program, those who want to understand the type of information that is covered in an MBA program, and those who would benefit from having an overview of the content of a typical MBA program, for example. The speakers are subject matter experts who will take into account the interests and background of chemical engineers in the way that they present the information. The average student who is enrolled in an MBA program has six years of experience and is 29 years old. Students with engineering or science backgrounds comprise 35 to 40% of MBA programs. As a consequence, the session will be highly of interest to young professionals

(43a) Looking to the Future in Continuous Pharmaceutical Manufacturing

Richard D. Braatz, Chemical Engineering, Massachusetts Institute of Technology, Cambridge, MA

This presentation describes the motivations, challenges, and opportunities in continuous pharmaceutical manufacturing. Some of the motivations are particularly intriguing, such as semicontinuous and continuous-flow unit operations that have the potential to have order-of-magnitude higher controllability and/or are much easier to scale up than batch and semibatch processes. Theoretical, computational, and experimental results are provided as evidence of the potential opportunities of continuous-flow pharmaceutical manufacturing operation. One of the conclusions of reviewing such unit operations is that they can look and operate very different from the batch and semibatch unit operations currently used in the pharmaceutical industry.

Example unit operations are drawn from the published literature, which include both journal papers and patents from various academic and industrial research groups. The presentation will also review and comment on the application of Quality-by-Design methodologies to pharmaceutical manufacturing. Of particular interest is the generation of design spaces, and the pros and cons of different approaches. In particular, it is shown that many of the methods for the construction of design spaces that have been proposed in the literature are useless from a control or operations point of view.

(43b) New Challenges and Opportunities for Pharmaceutical Manufacturing Science

Vincent Vilker, Office of Testing and Research, FDA, Chevy Chase, MD and Mansoor A. Khan, Cder/Ops/Otr/Dpqr, Food and Drug Administratin, Silver Spring, MD

New Challenges and Opportunities for Pharmaceutical Manufacturing Science

Vincent L Vilker and Mansoor A Khan, CDER/FDA

The increasing sophistication of modern drug products is leading to unprecedented challenges and opportunities for pharmaceutical manufacturing. At the same time, scientific and technological advancements in analytical instrumentation, coupled with implementation of the principles embedded in quality by design, enable the improved process monitoring and control that lead to better and more consistent product quality. ICH Q8(R2), the first of three International Conference on Harmonization (ICH) Guidance for Industry that spell out the concepts of Quality by Design, states that critical formulation attributes and process parameters should be identified through an assessment of the extent to which their variation can have impact on the quality of the drug product. In this presentation, examples will be given where modern instrumentation and manufacturing practices are required to maintain the critical formulation attributes of several drug products that have unique physicochemical and/or biological properties. Examples may include properties such as very low solubility, polymorphic solid states, poorly defined macromolecular size distributions, or where there is a tendency to become unstable in the formulated product.

() Recent Advances in the Molecular Modeling Foundation of Systems-Based Pharmaceutics

Constantinos C. Pantelides, Department of Chemical Engineering, Imperial College London,Center for Process Systems Engineering, London, United Kingdom

There are potentially significant benefits from the adoption of a systems-based approach to pharmaceutical science and engineering based on detailed mathematical first-principles models. However, the success of such an approach is predicated on the accurate description of the thermodynamic, kinetic and mechanical properties of the materials involved.

In this presentation, we review some recent advances in the description of the solid and fluid phases arising from work by the Molecular Systems Engineering group at Imperial College London. In particular, we review the state of the art for the *ab initio* prediction of crystal structure and polymorphism, and the prediction of vapour/liquid and liquid/solid phase equilibria using the SAFT-gamma group-contribution based equation of state. Significant progress has been achieved in all these areas over the last few years, but there are also significant challenges that still need to be addressed.

(43d) Experience with Regulatory Issues for Biomedical Devices and Vision for the Use of Advanced Control In Pharmaceutical Operations

Francis J. Doyle III, Chemical Engineering, UCSB, Santa Barbara, CA, Eyal Dassau, UCSB, Santa Barbara, CA and Howard Zisser, Sansum Diabetes Research Institute, Santa Barbara, CA

The landscape for clinical testing of biomedical devices has evolved rapidly over the last 5-10 years. In particular, the use of advanced control algorithms, such as model predictive control, are now being reviewed by the FDA as components of closed-loop devices. In our work with the Artificial Pancreas (AP), we have developed master files and other supporting materials to support IDE (Investigational Device Exemption) studies. In this talk, I will share some of the challenges in developing algorithms for regulatory review, including the role of simulation studies, communication architectures, algorithmic design, and IDE studies. I will extrapolate from these experiences to project the opportunities for advanced control in pharmaceutical operations.

Chemical Engineering & Biopharmaceutics: Linking the Process with the Ultimate Quality Attribute

Ravi M. Shanker

The discovery and development of heterocyclic, small molecule new drug entities is a highly sophisticated, multi-disciplinary, scientifically advanced, collaborative, highly competitive, extremely expensive and multi-year endeavor. The heavily R&D dependent pharmaceutical industry utilizes cutting edge scientific and technological tools and techniques encompassing the broad disciplines of chemistry, biology and medicine. To provide significantly improved therapeutic treatments for various acute and chronic diseases and to be able to meet the requirements and expectations of the regulatory agencies, the patients and the society is a constant challenge for the innovative pharmaceutical industry. The challenges forge deeper dives in science among the wide array of disciplines that constitute the pharmaceutical industry and consequently have provided tremendous opportunities for the advancement of computational and predictive tools within each of these disciplines and sub-disciplines. A key segment of R&D and industrial scale manufacture is pharmaceutical sciences. It is an amalgamation of applied disciplines of chemical engineering, physical and physical organic chemistry, synthetic organic chemistry, process chemistry, analytical chemistry, polymer chemistry, material sciences, crystallization science, drug product design and manufacture and evaluation of performance of drug product in the patients. Pharmaceutical sciences have also embraced the development and use of predictive computational tools among each of its areas of specialization and disciplines to ensure the highest quality of drug product being delivered to the patient. The multi-disciplinary components of pharmaceutical sciences are operationally integrated to provide the highest quality of drug product to meet the regulatory requirements in this highly regulated industry. However, there is a lack of integration of such predictive tools across disciplines to enable the assessment of local perturbations in quality during any segment of work flow on the ultimate quality and performance of the drug product. The ultimate quality attribute of a drug product is it's performance in patients, however, linking that performance to the upstream attributes of drug substance and drug product via engineering control during manufacture with emphasis on the principles of chemical engineering continues to be unrealized and unfulfilled possibility. The vision to provide such integrated approach of existing computational and predictive tools within disciplines and sub-disciplines in pharmaceutical sciences and to provide opportunities for further development of advanced integrated tools as a holistic system is the underlying concept for the creation of Systems Based Pharmaceutics.

Large-Scale Fabrication of Polymer-Clay Nanocomposites with Effective Exfoliation and Dispersion by Solid-State/Melt Extrusion (SSME)

Alyssa M. Whittington [a], Stephen M. Brouse [a], Michael A. Malusis [b], and Katsuyuki Wakabayashi [a,*]

[a] Department of Chemical Engineering, Bucknell University, Lewisburg, PA 17837, USA

[b] Department of Civil and Environmental Engineering, Bucknell University, Lewisburg, PA 17837, USA

*Corresponding author. Tel.: +1 570 577 3778; Fax: +1 570 577 1141; E-mail: kw025@bucknell.edu

Abstract

In this communication, solid-state/melt extrusion (SSME) is introduced as a novel technique that combines solid-state shear pulverization (SSSP) and conventional twin screw extrusion (TSE) in a single extrusion system. The morphology and property enhancements in a model linear low density polyethylene/organically modified clay nanocomposite sample fabricated via SSME were compared to those fabricated via SSSP and TSE. The results show that SSME is capable of exfoliating and dispersing the nanofillers similarly to SSSP, while achieving desirable output rate and shape similar to those of TSE.

(59b) Magnetic Epoxy Resin Nanocomposites Reinforced with Fe@C Nanoparticles

Ouassima Alloul[1], Jiahua Zhu[1], Suying Wei[2] and Zhanhu Guo[1], (1)Integrated Composites Laboratory (ICL), Dan F Smith Department of Chemical Engineering, Lamar University, Beaumont, TX, (2)Chemistry and Biochemistry, Lamar University, Beaumont, TX

Magnetic epoxy nanocomposites comprising carbon coated iron (Fe@C) nanoparticles have been produced in this work. Fe@C nanoparticles were suspended in epoxy resin monomer to investigate the rheological properties and the corresponding cured epoxy nanocomposites are fabricated with a heating treatment. Effects of particle loading on the rheological properties, including complex viscosity, storage modulus and loss modulus are systematically investigated. Unique rheological phenomena "a decreased viscosity with a better dispersion" are observed and explained in terms of the improved filler dispersion quality. Meanwhile, significant increased storage modulus in cured nanocomposites is observed and related to the better dispersion and strong interfacial interaction between NPs and epoxy matrix as revealed by SEM imaging. The Fe@C nanoparticles become magnetically harder after dispersing in epoxy than that of pure Fe@C nanoparticles due to the decreased interparticle dipolar interaction, which arises from the enlarged NP spacer distance for the single domain nanoparticles. Finally, electrical resistivity measurements indicate a threshold percolation of the nanoparticles between the loading of 5 wt% and 10 wt%, where a significant reduction of resistivity is observed.

(59c) Hollow Fe@SiO2 Nanorods for Oxidation-Reduction Chemistry

Hannah Grace, Chemical Engineering, University of Pittsburgh, Pittsburgh, PA, Michelle Najera, Department of Chemical Engineering, University of Pittsburgh, Mascaro Center for Sustainable Innovation, University of Pittsburgh, Pittsburgh, PA and Götz Veser, Department of Chemical and Petroleum Engineering, University of Pittsburgh, Pittsburgh, PA

Core-shell materials, consisting of a reactive metal core surrounded by an inert support shell are of interest in a range of applications including catalysis, drug delivery, optics, and separation processes. *Hollow* core-shell materials, in which there exists a central cavity within the metal core, are of particular interest; they present new possibilities for limiting confinement effects and hence studying the potential for an enhanced reduction-oxidation capacity for the metal cores. We have developed a synthesis path for hollow iron core/silica shell nanoparticles (Fe@SiO_2) with ~40 x 150 nm rod shaped hollow iron cores. The method begins with a hydrothermal synthesis of β-FeO(OH) nanorods, followed by a wrapping of these particles with a 20-30 nm porous and amorphous silica shell via a modified Stöber synthesis. This synthesis route relies on the low density of the β-FeO(OH) nanorod intermediate. The metal core becomes hollow only after transformation into higher density iron materials via calcination in air (to Fe_2O_3) or reduction (to elemental iron). We have found that although a range of both hydrothermal and forced hydrolysis synthesis conditions can lead to the β-FeO(OH) intermediate, an optimization of temperature, iron precursor concentration, and pH are required for well-defined and uniform nanorods with narrow particle size distributions.

Notably, initial evaluation of this material indicates redox capacities ~2 times higher than conventional (non-hollow) Fe@SiO_2 materials, possibly due to reduced confinement effects. Synthesis and characterization of the materials, as well as application as oxygen carriers in chemical looping, a clean combustion technology, will be discussed in detail in the presentation.

(59d) Nanoconfined Homogeneous Catalysts for Acylation Reactions

Martin Roberts, Chemical Engineering Department, University of Pittsburgh, Pittsburgh, PA, David Palmer, University of Pittsburgh, Pittsburgh, PA, Karen J. Uffalussy, Chemical and Petroleum Engineering, University of Pittsburgh, Pittsburgh, PA, Xinyu Liu, Department of Chemistry, University of Pittsburgh, Pittsburgh, PA and Götz Veser, Department of Chemical and Petroleum Engineering, University of Pittsburgh, Pittsburgh, PA

Homogeneous catalysts are typically much more selective than heterogeneous ones, but used much less frequently in industry due to the difficult separation of the catalyst from the reactant/product mixture and hence limited re-use of catalyst. Hence, immobilization of homogeneous catalysts has become the focus of broad efforts over recent decades with the goal of facilitating separation of the catalyst and its repeated re-use without loss of activity. Among the different approaches toward immobilization, in particular adsorption, tethering (or grafting), and, more recently, encapsulation have been actively pursued. Both adsorption and tethering often result in configurational changes of the catalyst molecule which can negatively affect its catalytic activity. Encapsulation, on the other hand, avoids this issue but often requires complex (and expensive) preparation procedures.

In the present work, we are investigating the encapsulation of homogenous acylation catalysts in inorganic core-shell nanostructures. Specifically, pyridine and 4-dimethylaminopyridine (DMAP), two well-established acylation catalysts were encapsulated in silica "nanobubbles", i.e. hollow silica nanoparticles (~30 nm diameter), in which silica walls (~10 nm thickness) enclose a pronounced central cavity. The walls are highly porous allowing for facile reactant access to the central cavity. The materials were synthesized and characterized (TEM, BET), loaded with pyridine or DMAP, respectively, through a simple wet impregnation procedure, and then tested in the acylation of 4-methoxybenzyl alcohol with acetic anhydride to 4-methoxybenzyl acetate. Thin layer chromatography (TLC) was used as a quick and simple method to monitor the reaction progress. Our results to-date demonstrate that it is indeed possible to encapsulate the homogeneous catalysts through this simple preparation procedure, and that the encapsulated catalysts maintained their catalytic activity. Further detailed studies of the impact of the nano-encapsulation on the kinetics of the reaction, as well as recyclability of the encapsulated catalysts are currently under way.

Overall, our results indicate that the encapsulation of homogeneous acylation catalysts into such silica "nanobubbles" could be a simple and promising route for immobilization of this class of catalysts.

(59e) Exergetic Analysis of Chemical Looping Reforming

Neha Nandakumar, Carnegie Mellon University, Pittsburgh, PA, Michelle Najera, Department of Chemical Engineering, University of Pittsburgh, Mascaro Center for Sustainable Innovation, University of Pittsburgh, Pittsburgh, PA, John R Kitchin, Chemical Engineering, Carnegie Mellon University, Pittsburgh, PA and Götz Veser, Department of Chemical and Petroleum Engineering, University of Pittsburgh, Pittsburgh, PA

Chemical looping combustion (CLC) is an emerging clean combustion technology in which metal oxygen carriers are exposed to oxidation/reduction cycles with air and fuel, respectively. The oxygen carrier is first oxidized by air in the oxidizer, forming a metal oxide, and then reduced by a fuel in the reducer, forming CO_2, water, and the reduced form of the metal. Overall, CLC yields a flameless, NOx-free combustion and produces sequestration-ready CO_2. CLC has long been recognized as a highly efficient energy process when compared to conventional combustion. This increased energy efficiency is achieved via the division of the conventional gas phase combustion reaction into the two reversible gas-solid reactions in the CLC oxidizer and reducer. The high efficiency is due to a reduction in the entropic losses, or irreversibilities, associated with the combustion process, i.e. chemical looping has high exergetic efficiency, resulting in more usable energy when compared to conventional gas phase combustion.

Beyond CLC, several chemical looping reforming (CLR) technologies have also been identified. For instance, the use of steam instead of air as the oxidant has been demonstrated by our group and others to yield an efficient route produce high-purity H_2 as the oxidizer effluent. Similarly, our group has recently proposed the use of CO_2 instead of air as the oxidant, resulting in chemical looping dry reforming (CLDR) as an novel method to utilize CO_2 via reduction to CO (i.e. via "CO_2 activation"). While the thermodynamic and kinetic feasibility was demonstrated for these chemical looping reforming processes, their exergetic efficiency has not been studied to-date.

In the present work, we have developed a process model for CLR processes and implemented it in the Aspen modeling software. First results from these studies to-date indicate an increased exergetic efficiency of chemical looping dry reforming (CLDR) compared to conventional dry reforming. Extension onto H_2 production via CLR processes and comparison to conventional H_2 production via steam reforming and partial oxidation is currently under way.

(59f) Transport Properties of Methane Confined in Nanoporous Carbons

Andrew P. Santos, Chemical and Biomolecular Engineering, NC State University, Raleigh, NC, Joshua D. Moore, Weapons and Materials Research Directorate, U.S. Army Research Laboratory, Aberdeen Proving Ground, MD, Jeremy C. Palmer, Department of Chemical and Biological Engineering, Princeton University, New Jersey , NJ and Keith E. Gubbins, Chemical Engineering, North Carolina State University, Raleigh, NC

Nanoporous carbons are used widely in industry as adsorbents for gas- and liquid-phase separations due to their excellent surface properties, which arise from their high specific surface area, porosity and nano-scale pores. Although significant efforts have been made to optimize the adsorptive properties of these materials for separations, the transport properties of guest phases confined within nanoporous carbons have received less attention. Using molecular simulation techniques, we have systematically examined the diffusive behavior of confined methane molecules inside a model nanoporous carbon in order to understand how the behavior is influenced by structural characteristics such as pore size, tortuosity, and pore connectivity. We report results from our study, beginning with our analysis of high pressure adsorption isotherms of methane calculated using Grand Canonical Monte Carlo simulations. Self, collective and transport diffusion coefficients calculated using microcanonical molecular dynamics are also reported at different state conditions along the isotherms. We find that these transport coefficients have a non-monotonic dependence on the loading and bulk pressure, indicating that the rate of transport is significantly altered by the state conditions. An analysis of first passage and mean survival times is also presented in order to examine how methane molecules are transported between spatially segregated regions within the heterogeneous porous network. Finally, we show that there are significant differences in these properties depending on whether methane is described using a united-atom or all-atom force field, suggesting that the latter is more appropriate for studying methane in confinement.

(61a) Chemical Engineering and Kinetics: a "Pas de Deux" of theory and experiment

Guy B. Marin, Laboratory for Chemical Technology, Ghent University, Ghent, Belgium

Chemical kinetics constitute a key discipline for the chemical engineering profession. They allow the design of new and the optimization of existing products and processes. This is not only so because they provide quantitative relations between process input and output but even more importantly because they provide insight in the reaction mechanism underlying these relations.

The tremendous progress made during the past decades in the field of computational chemistry and engineering has lead the profession away from the classical road from concept to industrial realization via a set experimental data collection campaigns at increasing scales. Product and process innovation now requires a much more sophisticated interaction between theory and experiment. There is, of course, the "straightforward" quantum chemical generation of reaction networks and corresponding rate coefficients and of catalyst or polymer properties. Experimental verification of such data remains a must for the decades to come. More interesting, because requiring heuristics i.e. the intuition of the researcher, are cases where theory provides information which is quasi impossible to obtain experimentally but leads to new or further insights. Less extreme is the case that tools such as Computational Fluid Dynamics (CFD) codes allow to provide the proof-of-concept required before engaging into an experimental program. Finally there is a caveat: any theory starts from premises and, hence colors the insights it provides. So-called model-free analysis of experimental data is aimed at avoiding this pitfall.

This lecture will illustrate this "Pas de Deux" between theory and experiment in polymerization, catalysis and reactor engineering. This will provide the opportunity to discuss theoretical and experimental techniques such as kinetic Monte-Carlo and CFD simulations and Temporal Analysis of Products (TAP) experimental data. The latter will be described in terms of so-called microkinetics i.e. accounting for the relevant elementary reactions and highlighting the role of the catalyst.

(63a) CO_2 Capture, Utilization and Storage and Natural Gas/Oil Recovery – Harnessing Scientific Development and Business Principles to Achieve Fossil Energy Sustainability

Charles McConnell, U.S. Department of Energy (DOE), Washington, DC

Fossil fuels currently comprise more than 80 percent of global (and U.S.) energy consumption and are expected to continue to play a critical role in meeting the world's energy needs for the foreseeable future. Making use of these fuel sources in a cost-effective and environmentally sustainable manner will require technologies that can be broadly applied in the marketplace. The Department of Energy's Office of Fossil Energy is focused on the research, development and deployment (RD&D) of technologies that combine the latest scientific and technical advances with sound business principles to achieve fossil energy sustainability in an economically viable manner. Key examples of this RD&D include carbon capture, utilization and storage (CCUS) technologies to achieve carbon dioxide (CO_2) mitigation while storing and commercially utilizing the captured CO_2, and advanced approaches to natural gas and oil recovery, such as hydraulic fracturing. Going forward, a critical focus will also be on ensuring the safety and sustainability of these technologies and practices.

(63b) Comparative Advantage - North American Manufacturing and the Shale Gas Century

David Porges, EQT Corporation, Pittsburgh, PA

Over the long term, there are two fundamentals in North America's natural gas markets:
1. Supply will remain robust; and
2. Price will remain moderate and stable.

This new-found, but likely durable, situation provides many opportunities to resuscitate the petrochemical industry in North America. The challenge is also twofold: the natural gas production business must develop this asset in the most sustainable manner, and the petrochemical industry must determine how best take advantage of this ample, inexpensive fuel and feedstock.

(63c) Chemistry and Energy: Fortifying Our Historic Links

Gregory Babe, Orbital Engineering, Pittsburgh, PA

Centuries ago, early ideas were formulating that would advance the science of organic chemistry. Since its beginnings, there have been many organic chemistry milestones. They include in 1856, the discovery of a purple dye which helped usher in the modern organic chemistry industry.

Henry Perkin, a precocious 18-year-old British chemist, synthesized the first synthetic dyestuff from aniline, a chemical molecule of coal tar. Perkin's discovery was accidental. He had set out to produce quinine, a drug to treat malaria.

From this lucky accident grew the synthetic dyestuffs industry. In 1857, with financial help from his father, Perkin opened a factory near London to produce his new purple dye. In Barmen Germany, just six years later, Friedrich Bayer and Johann Weskott began making dyestuffs from coal tar derivatives in the Weskott family sink—a very modest beginning for today's Bayer Group. Yet without the proliferation of gas lighting in the 19th century, there would have been be no readily available coal tar derivatives. The synthetics dyestuffs that allowed the then young chemistry industry to flourish would not have been possible.

Chemistry's historic links to energy production are often overlooked. Yet the industry's raw materials still come from the same hydrocarbons that made possible the Industrial Revolution and mass production miracles of the last century.

Oil—and especially natural gas—continue to be the chemistry industry's building blocks for more than 96 percent of all manufactured goods. Yet the industry has been whipsawed over the years by natural gas supply shortages and high prices. Between 1999 and 2005 alone, U.S. natural gas prices quadrupled. Consequently, the U.S. chemistry industry shed more than 120,000 jobs in the last two decades. Many of those jobs were lost to nations with lower natural gas prices.

Lower natural gas prices and increase supplies from shale gas can be a game changer for the chemistry industry. An American Chemistry Council study concludes that American shale gas development can lead to 400,000 chemistry and supply industry jobs, $16 billion in capital investment, $132 billion in economic output, and more than $4 billion in federal, state and local tax revenue each year. The benefits are already beginning to roll in.

What's more, shale gas can yield ethane, an essential petrochemical feedstock. The Appalachian States are prime locations for new petrochemical facilities. An ACC study found that a new petrochemical facility in Ohio would yield nearly $8 billion in new output, 17,000 chemistry and supplier jobs, $1 billion in wages and nearly $170 million in state tax revenue annually.

Last year, a TIME magazine cover story on shale gas reported, "This Rock Could Power the World." That's hyperbole. But there is no denying that we've entered a golden age of natural gas supply.

America needs an "all of the above" energy strategy. We need one that promotes all domestic energy resources, from coal, oil and natural gas to nuclear power and renewables. We also need to rigorously promote energy efficiency. In this regard, chemistry companies are leading the way in developing energy-efficient products and technology. Examples include lightweight automotive products, building insulation, and materials and technologies used in solar panels, wind turbines and lithium batteries.

To get to where we need to be, industry must work to remove four obstacles that stand in the way of success. They are communication, politics, behavior and science education.

Most people don't understand the upsides of the chemistry and energy industries, and the strong links that form fundamental economic strengths for our nation. We need to start a conversation and keep it going with opinion leaders, elected officials, regulators, NGOs, plant communities, critics and employees. We must listen to them, act on what we hear and use every communication channel available.

Paradoxically, even as public trust of government is diminishing, legislation—some of it ill-conceived— is growing. Far too often, the chemistry and energy industries are at the end of the legislative pipeline. We should be at the beginning, designing solutions for the future instead of fighting yesterday's battles. We must continuously advance the understanding of how our operations affect human health and the environment. Only then can we ensure that legislation is based on sound science.

The chemistry and energy industries must take all safety, facility security and environmental protection processes to a higher level. We must hold ourselves to the highest standards, admit where we have fallen short and follow up with corrective action.

On the education front, The Department of Education estimates that America will need 400,000 new graduates in STEM fields by 2015. We're not likely to get them. America's students continue to lag in science education at a time when they need it to succeed in a highly competitive, worldwide economy. We must reverse this trend and begin with students at an early age.

The chemistry and energy industries have come a long way together since Henry Perkin's accidental discovery. We have a lot in common. Working together, we can break down the barriers to continued progress.

(63d) Nuclear Energy: Is there a future after Fukushima?

Aris Candris, Westinghouse Electric Company, Monroeville, PA

The future of the global nuclear energy industry, while certainly not without its challenges, remains vibrant. Even as the industry contends with near-record-low natural gas prices and the accident at Fukushima Daiichi, the reality is that global demand for electricity and concern over carbon emissions will ensure that nuclear energy becomes an even larger source of global electricity. In fact, as a result of the accident at the Fukushima Daiichi nuclear power plant, the passive safety design of the Westinghouse AP1000 nuclear power plant is making it an even more popular choice in the global new plant market. Additionally, Westinghouse is working on next-generation nuclear technology that extends the benefits of the AP1000.

ENVIRONMENTAL & CARBON REPORTING & DISCLOSURE:

NOT BUSINESS AS USUAL

By

John. P. Fillo, Ph.D. CPEA

EXTENDED ABSTRACT

Climate change and carbon risk management are complex environmental and societal issues, as well as being fundamental to business sustainability. Increased regulation and reporting of environmental and carbon exposures, liabilities and risks are driven by national, regional, state/provincial and local requirements in the U.S. and abroad. These obligations have evolved from participation in voluntary initiatives to legislative and regulatory-driven mandatory actions. While reporting of greenhouse gas (GHG) emissions is only one component of these requirements, statutory financial disclosure requirements for publicly held companies and even country-specific requirements for environmental/ sustainability reporting extend these obligations to address strategy, operations, business risk and opportunity management, legal compliance and supply chain issues, as well as disclosures in financial statements and reports.

This presentation will provide an overview of and insight into the complexity of these requirements and their relevance to the chemical engineering profession. While it will provide a brief overview of the mandatory 'environmental' reporting programs and requirements as a baseline, the focus will be on statutory, industry and voluntary environmental, carbon and sustainability reporting and disclosure in the U.S. and internationally. These requirements and voluntary venues represent complex carbon reporting and disclosure requirements, and can extend beyond just simply reporting a GHG inventory. Such disclosures often require that corporate strategy, risk and opportunity management, and programs and initiatives be disclosed as a basis for assessing the veracity of such disclosures.

Mandatory Reporting & Disclosure

The U.S. is faced with a patchwork quilt of GHG reporting and disclosure requirements at the Federal, regional and state levels:

- The U.S. EPA's GHG Mandatory Reporting Rule, for which annual reporting of Scope 1 GHG emissions is required for numerous industry sectors.

- Regional GHG programs are a work in progress, with the most active being the Regional GHG Initiative (RGGI), the first market-based regulatory program in the U.S. to reduce GHG emissions.
- Several states have or are developing GHG reporting and regulatory programs, with California's Global Warming Solutions Act (AB-32) enacted, with GHG inventory reporting and verification requirements in place and cap and trade program requirements recently issued in 2012.

Additionally, U.S. Federal sustainability and carbon planning and reporting requirements have been in place since late 2009 under Executive Order 13514, *Federal Leadership in Environmental, Energy and Economic Performance*. The Executive Order requires annual reporting of Scope 1, 2 and 3 GHG emissions, as well as setting and achieving goals for energy use and GHG emissions reductions, and improvement of broad-based sustainability performance.

And our neighbors to the North have GHG reporting requirements, as well. Canada has Federal annual GHG emissions reporting requirements. Several provinces require industry-wide reporting of GHG emissions (e.g., Alberta, British Columbia, Quebec and Ontario), with some requiring formal verification of these emissions above stipulated thresholds.

Environmental, social and carbon reporting and disclosure requirements are typically embedded in securities reporting regimes throughout the world:

- Environmental and carbon risk reporting and disclosure are mandated by the U.S. Securities and Exchange Commission under Regulation S-K. Similar statutory requirements exist in Canada through Provincial securities commission requirements.
- European companies are required to include non-financial information in their annual and consolidated reports per the European Accounts Modernization Directive, and country-specific enabling legislation that can include requirements for reporting and disclosure of environmental, social and governance information.
- Country-specific requirements can mandate environmental and/or sustainability reporting (e.g., Japan, Brazil).

And while in the early stages of development, additional climate change and sustainability related reporting and disclosure mechanisms are being developed, which have the potential to become mandatory in the future:

- Global carbon disclosure and attestation standards, including those from the Climate Disclosure Standards Board, provide guidance on carbon risk disclosure for companies that are expected to follow International Financial Reporting Standards.

- The International Integrated Reporting Council is leading the development of a global framework for corporate reporting that demonstrates the linkages between an organization's strategy, governance and financial performance, and the social, environmental and economic context within which it operates.
- The Sustainability Accounting Standards Board, a US based, non-profit organization, intends to establish industry-based sustainability standards for the recognition and disclosure of 'material' environmental, social and governance impacts by companies traded on U.S. stock exchanges.

Voluntary/Industry Programs

While mandatory programs developed over the past few years in the U.S., GHG reporting and disclosure has been driven by a variety of voluntary programs and initiatives. These include, but are not limited to:

- The Carbon Disclosure Project, an independent not-for-profit organization, headquartered in the UK that seeks disclosure on climate change business risks and opportunities and GHG data from the world's largest companies since 2000. Reporting programs have extended into the areas of supply chain, water, and a cities project, as well.
- The Climate Registry, established in North America to measure, report and verify GHG emissions consistently across industry sectors.
- The Carbon Principles is a collaboration between investor-owned utilities, banks and NGOs to provide a consistent approach for banks and their U.S. power clients to evaluate and address carbon risks in financing electric power projects.
- The National Association of Insurance Commissioners has developed a reporting and disclosure framework that applies to insurers with aggregate annual premiums over $300 million, who are required to submit annual climate risk reports.

And there are additional voluntary venues, including GHG mitigation project development through carbon offset registries (e.g., Climate Action Reserve, Verified Carbon Standard, and others) and industry trade organization programs (e.g., Global Environmental Management Initiative, Asia-Pacific Partnership on Clean Development and Energy, and others).

Sustainability Indices

Sustainability indices can be an important venue through which to achieve reporting, disclosure and recognition. While these are frequently voluntary, representativeness, completeness, consistency, accuracy, and balance of disclosures speaks to corporate credibility and a company may be evaluated even if you choose not to formally participate. Corporate sustainability reporting is typically one of many criteria utilized to rank the included companies. Such indices include but are not limited to the Dow Jones Sustainability Index, FTSE4Good Index, Calvert Social Index, Morgan Stanley KLD 400 Social Index, the Newsweek Green Rankings and others.

Implications

And are these relevant to the chemical engineering profession? They most certainly are. For a chemical engineer whose career has evolved down a path of EHS/sustainability management or into the legal profession, they will be directly involved in the reporting and disclosure process. For the chemical engineer who is focused to product/process R&D or design and engineering, understanding these issues is essential to assure they are addressed progressively through the process and product development lifecycle. Engineering efficiencies and pollution prevention into processes and products, and assuring continuous improvement, is essential so as to improve a corporation's top and bottom line, and to ensure that the necessary information is available to support these required disclosures.

(77b) Epa's Clean Air Act Authority to Regulate Greenhouse Gases

Mary Ellen Ternes, McAfee& Taft, Oklahoma City, OK

Ever since the passage of the 1990 Clean Air Act Amendments, regulation of greenhouse gases has been controversial. Throughout the George W. Bush administration, EPA creatively avoided adopting Clean Air Act regulations governing greenhouse gases. Then, with the 2007 landmark United States Supreme Court case, Massachusetts v. EPA, the EPA was required to directly address greenhouse gas emission sources pursuant to the Clean Air Act. In 2009 and 2010, EPA initiated several rulemakings governing greenhouse gas emissions, including the Greenhouse Gas Reporting Rule, the Endangerment and Cause and Contribute Finding, the mobile source emission standards (the "Tailpipe" rule), the Tailoring and Timing Rules. Throughout this period, the U.S. House of Representatives entertained several proposals to eliminate EPA, combine EPA with the Department of Energy or simply modify the Clean Air Act to remove any authority to regulate greenhouse gas emissions. Then, in 2012, the D.C. Circuit reviewed all the legal challenges to the suite of greenhouse gas rulemakings and issued its decision. This presentation will review the D.C. Circuit's decision regarding the validity of EPA's greenhouse gas regulations, impacts from the decision, and implications for future policy.

(77c) Synthesis of CaO-Scaffold Adsorbent with Superior Stability for High-Temperature CO_2 Capture

Ming Zhao, School of Chemical & Biomolecular Engineering, The University of Sydney, Sydney, Australia, Nicholas H. Florin, Chemical Engineering, Imperial College, London, London, United Kingdom, Paul S Fennell, Chemical Engineering and Chemical Technology, Imperial College London, London, United Kingdom and Andrew T. Harris, Chemical Engineering, University of Sydney, Sydney, Australia

Introduction

Reducing greenhouse gas emissions to the atmosphere is a key environmental issue facing the world in general. Carbon dioxide (CO_2), primarily from the combustion of fossil fuels for energy production, is the most common greenhouse gas emitted by human activities and the one with the most significant impact on our climate. Calcium looping is one of the most promising technologies for both pre- and post-combustion CO_2 captures at a wide temperature range (400-900 °C) due to the high CO_2 carrying capacity and economical availability [1]. However, the natural CaO-based sorbents (limestone, calcite, dolomite, etc.) and even the synthetic species with tuned structures are suffering from a rapid decay in activity because of a serious sintering and fraction deduction, both of which happen to a pure CaO sorbent [2].

To solve the problem, we developed a sol-gel method to graft CaO onto an inert scaffold (silicon oxide) in order to mitigate sintering and enhance the mechanical properties. This method gives a new approach to a porous and stable CaO-based sorbent that would not only improve CO_2 capture but also enable the introduction of extra metallic sites to form a sorbent/catalyst composite.

Experimental and results

A saturated ethanol solution of tetraethoxysilane (TEOS) and $Ca(NO_3)_2$ was added dropwise into the mixture of 2.5 M ammonium hydroxide, hexadecyltrimethyl ammonium bromide (CTAB) and excess Na_2CO_3. The mixture was gelled and aged at 60 °C over a total of 24 h. The precipitate was washed, dried and calcined at 550 °C in air for 4 h. This CaO-scaffold sorbent was prepared with $n_{Si}:n_{Ca}$ = 1:1. A commercial sorbent that is stored as micrometer-sized $CaCO_3$ (Sigma-Aldrich) was used for comparison.

$$Si(OC_2H_5)_4(TEOS) + 2\,H_2O \xrightarrow{OH^{-1},\ \Delta} SiO_2\downarrow +4\,C_2H_5OH \quad (1)$$

$$Na_2CO_3 + Ca(NO_3)_2 \longrightarrow CaCO_3\downarrow +2NaNO_3 \quad (2)$$

CTAB was used as a template agent due to its positively charged molecules and formed growing micelles by combining negatively charged hydrated silica that derived from hydrolysis of TEOS [3]. At a higher temperature (60 °C) than conventional MCM-41 synthesis condition (25 °C), the aggregation and precipitation of SiO_2 (Eq. 1) was observed much accelerated so that a co-precipitation together with the formation of $CaCO_3$ (Eq. 2) was initiated. The resultant solid product was found with a one-fold framework instead of separately grown mixture. The morphology imaged by SEM is shown in Fig. 1 where numerous spherical nano granules (20-50 nm) can be observed with regularly arrangement simultaneously forming porosity in nano scale. Therefore, this CaO-scaffold sorbent gave a much higher surface area (111.1 m^2/g by BET model) than a pure-CaO-based sorbent (mostly <10 m^2/g). The local ESD spectra indicate an even distribution of Si and Ca that also reflects the one-fold framework. Free $CaCO_3$ phases are confirmed by powder XRD technique, which predicts a reactivity of capturing CO_2 molecules.

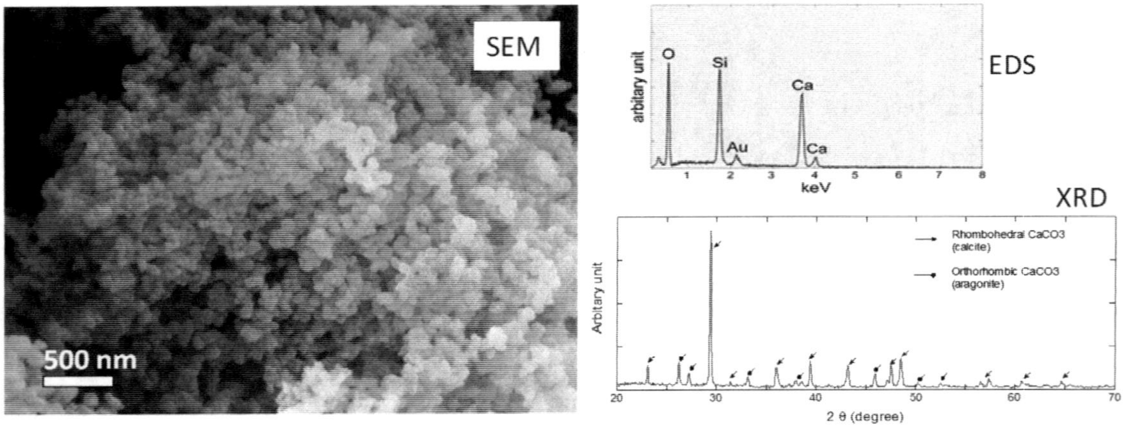

Fig. 1 Characterisation of the CaO-scaffold sorbent

The stability in multiple CO_2-capture-and-release runs was tested using a thermogravimetric analyser (TA Q5000IR). The sorbents were decarbonated in N_2 at 700 ^oC for 10 min followed by a cooling-down step to 600 ^oC. A 60-min carbonation in 60 vol.% CO_2 in N_2 then commenced. The decarbonation-carbonation process was repeated up to 13 cycles. The solid curves in Fig. 2 show the CO_2 carrying capacities (per unit sorbent) of the CaO-scaffold and commercial CaO sorbent, respectively, while the red dash line indicates the CO_2 uptake per unit CaO of CaO-scaffold, the value of which is different from the CO_2 uptake per unit sorbent due to the inert component. The new CaO-scaffold sorbent showed an excellent stability over multi-cycle tests despite the lower CO_2 uptake per unite sorbent compared to the commercial one. This stability reflected a superior ability to mitigate sintering deactivation that significantly decreases the practical potential of CaO-based CO_2 capture and is still investigated by current researchers. As shown by the red dash curve, the CO_2 binding capacity of free CaO on the new sorbent has surpassed the commercial one since the 2nd cycle and remained a nearly flat dropping track. This is likely attributed to the strong interaction between free Ca species with the scaffold [4].

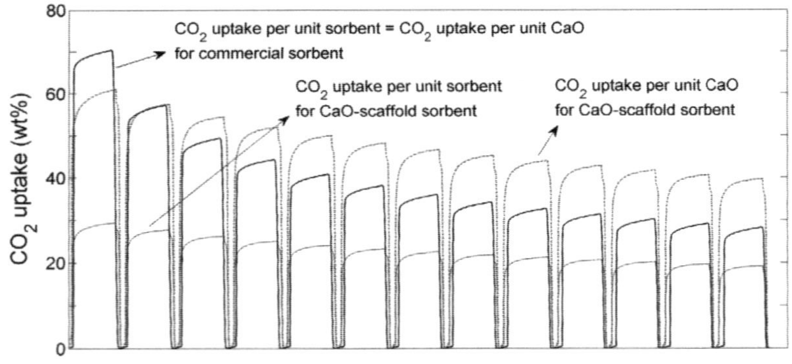

Fig. 2 CO_2-capture-and-release multiple tests using TGA

Conclusions

The demonstrated sol-gel method formed interaction between CaO and scaffold (SiO_2) and consequently a stable, porous, and ordered structure, which favored the sintering-resistant stability over multiple usages. Further investigations will be conducted for the deep understanding of the structure and optimization of the synthesis. The porous framework also provides space to house extra metallic catalyst, which would give a dual enhancement for hydrogen production from coal or biomass.

Acknowledgement. Authors are grateful to the Australian Research Council under DP0666488 and University of Sydney for International Program Development Fund.

References

[1]. Yang, Z.; Zhao, M.; Florin, N. H.; Harris, A. T., Synthesis and Characterization of CaO Nanopods for High Temperature CO_2 Capture. *Industrial & Engineering Chemistry Research* **2009,** *48*, (24), 10765-10770.

[2]. Florin, N. H.; Harris, A. T., Screening CaO-Based Sorbents for CO_2 Capture in Biomass Gasifiers. *Energy & Fuels* **2008,** *22*, (4), 2734-2742.

[3]. Cheng, C.-F.; Ho Park, D.; Klinowski, J., Optimal parameters for the synthesis of the mesoporous molecular sieve [Si]-MCM-41. *Journal of the Chemical Society, Faraday Transactions* **1997,** *93*, (1), 193-197.

[4]. Zhao, M.; Yang, X.; Church, T. L.; Harris, A. T., Novel CaO¨CSiO_2 Sorbent and Bifunctional Ni/Co¨CCaO/SiO_2 Complex for Selective H_2 Synthesis from Cellulose. *Environmental Science & Technology* **2012,** *46*, (5), 2976-2983.

In-Situ Synthesis of Useful Polyamines for CO_2 Capture from Piperazine

Omkar Namjoshi[1], *Yang Du*[1], *Han Li*[1,2], *Mark Goldman*[1], *Gary Rochelle*[1]

[1]The University of Texas at Austin, Austin, Texas, USA; [2]Tsinghua University, Beijing, PRC

Abstract

Concentrated piperazine (PZ) is considered to be a promising solvent for use in post-combustion CO_2 capture from coal-derived flue gas due to its fast rate of CO_2 absorption, high CO_2 capacity, and low degradation rates. However, this solvent has a limited solubility window at conditions corresponding to the operating temperature of a CO2 capture plant, limiting its use in commercial CO_2 capture systems. Blending PZ with a triamine analogue of PZ, such as aminoethylpiperazine (AEP), can widen the solubility window while maintaining many of the properties that make concentrated PZ attractive. With the exception of AEP, many of the triamine PZ analogues are expensive and are not commercially available. They can, however, be made in-situ in the regeneration section of the CO_2 capture plant by the selective reaction of PZ with an amine that forms an oxazolidinone (either a primary or secondary alkanolamine) as its initial thermal degradation byproduct.

The data suggests that the structure of the oxazolidinone is a strong function of reactivity with PZ. Increasing steric hindrance – such as increasing the chain length between the amino group or hydroxyl group or by adding alkyl functional groups to the carbon chain between the hydroxyl and amino groups – can reduce reaction rates by a factor of three to five. Secondary amines that form oxazolidinones react four times as fast as primary oxazolidinone-forming amines. Piperazine and solvent alkalinity losses are reduced significantly (~0.2%/week) once the oxazolidinone-forming amine is completely consumed, indicating that the synthesized amine blend is thermally stable.

Keywords: Carbon Capture, Amine Synthesis, Piperazine, Oxazolidinones

1. Introduction and Theory

A novel method for synthesizing triamines and tetramines in the regeneration section (stripping column) of an amine CO_2 capture plant was investigated. This process uses piperazine (PZ) or a PZ derivative and an amine whose initial degradation product is an oxazolidinone.

Oxazolidinones are the initial thermal degradation byproducts of primary or secondary alkanolamine with two or three carbon atoms between the oxygen and nitrogen atoms. They form when the amine carbamate (which, along with a protonated amine, is formed when CO_2 reacts with a primary or secondary amine) first undergoes electron abstraction followed by ring closing (Davis, 2009; Gouedard 2012). This scheme is shown in figure 1.1.

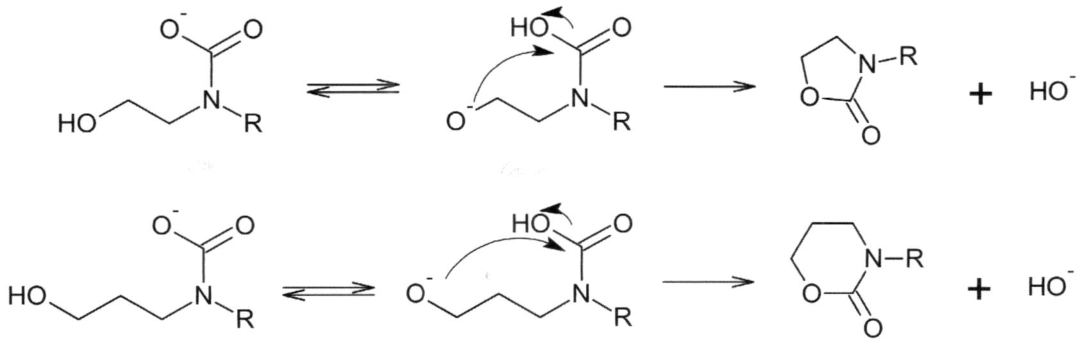

Figure 1.1. General scheme of oxazolidinone formation.

Primary and secondary amines can react rapidly with oxazolidinones to form substituted amine synthesis products. This general scheme is shown in figure 1.2.

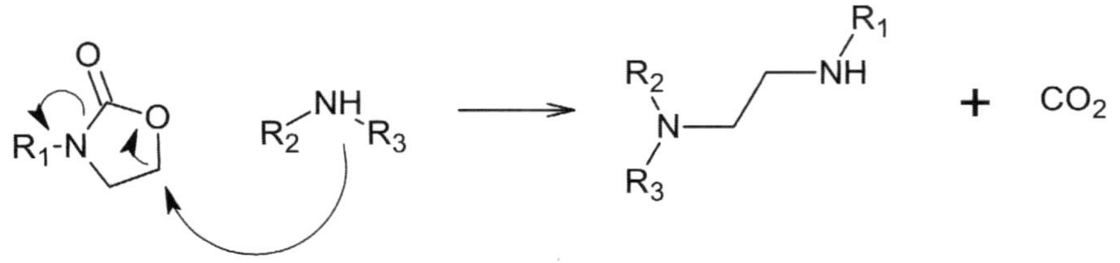

Figure 1.2. General scheme of reaction a primary or secondary amine with an oxazolidinone to form a synthesis product and CO2.

The oxazolidinone-forming amines studied in this paper include monoethanolamine (MEA), propanolamine (MPA), methylisopropanolamine (MIPA), 2-amino-2-methyl-1-propanol (AMP), and methylaminoethanol (MAE). Structures of these amines are shown in table 1.1. An example reaction scheme between PZ and MAE to form the synthesis product methylaminoethylpiperazine is shown in figure 1.3. The structures of the synthesis products formed by reaction with PZ are shown in table 1.2.

Table 1.1. List of the oxazolidinone forming amines used in this study

Amine Name	Abbreviation	Amine Structure	Structure of the Oxazolidinone
Monoethanolamine	MEA	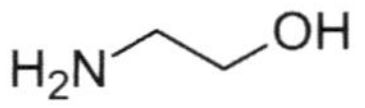	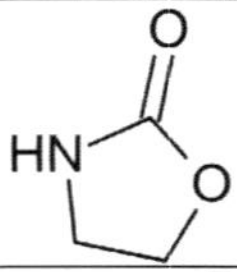
Methylaminoethanol	MAE	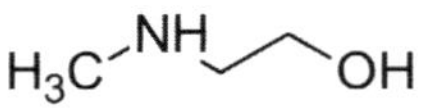	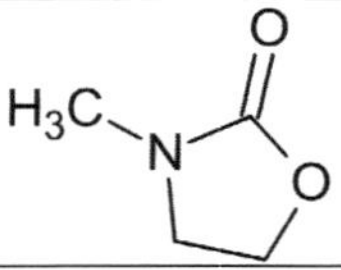
Monoisopropanolamine	MIPA	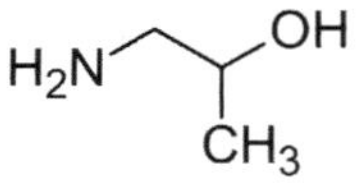	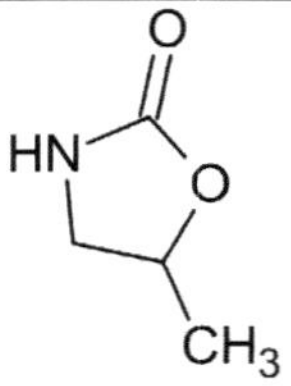
2-Amino-2-methyl-1-propanol	AMP	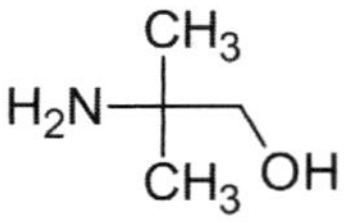	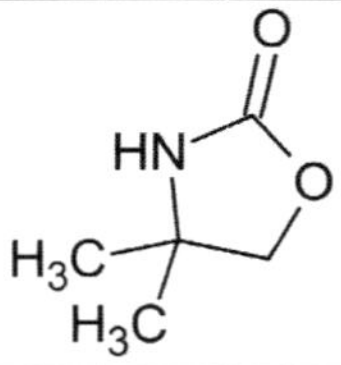
Monopropanolamine	MPA		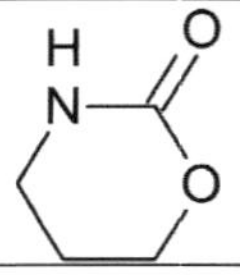

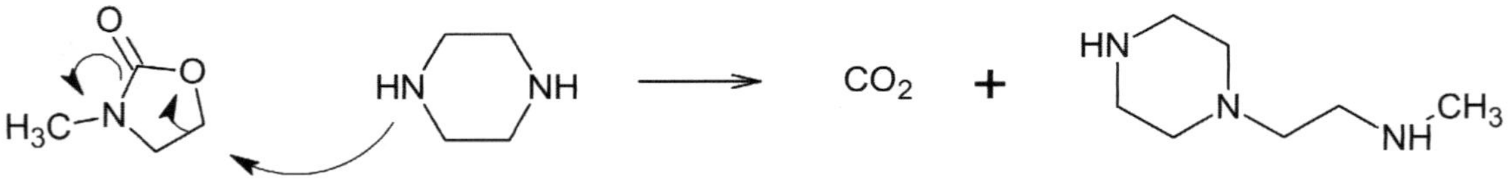

Figure 1.3. Formation of methylaminoethylpiperazine by reaction of PZ with MAE.

Table 1.2. Structures of the principal synthesis products seen in reaction between PZ and the oxazolidinone-forming amine

Amine	Triamine Synthesis Product	Tetramine Synthesis Product
Monoethanolamine	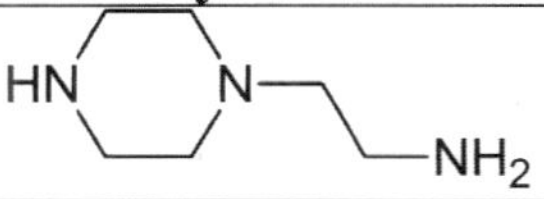	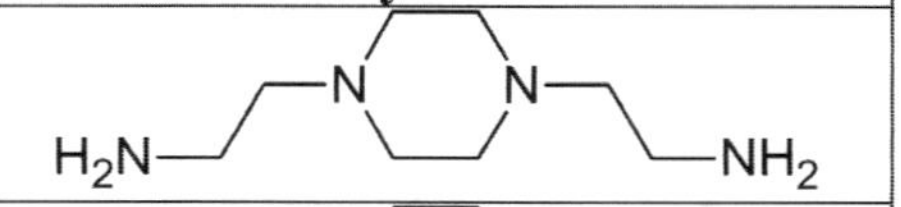
Methylaminoethanol	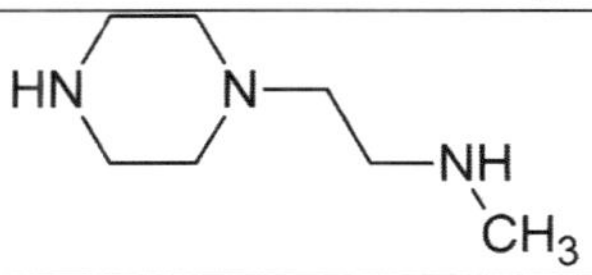	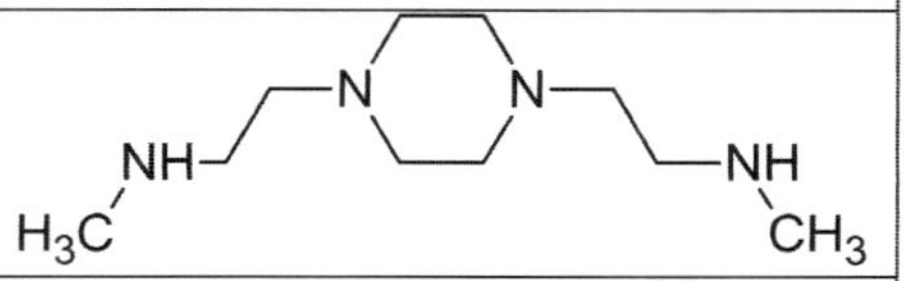
Monoisopropanolamine	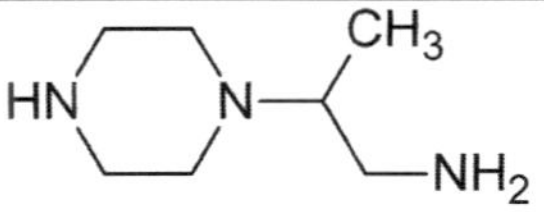	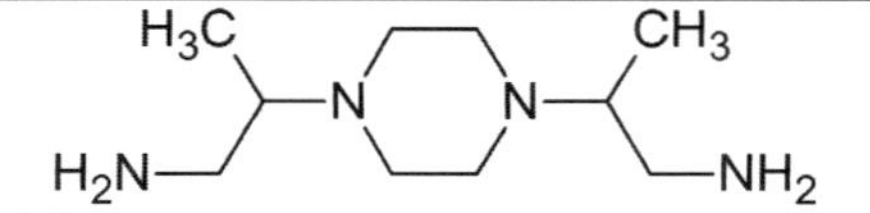
2-Amino-2-methyl-1-propanol	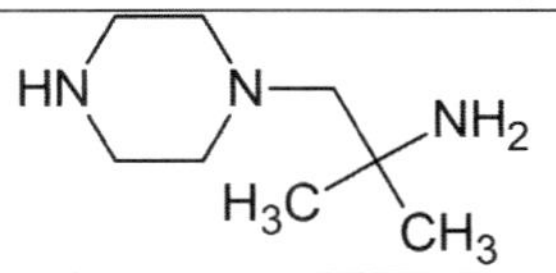	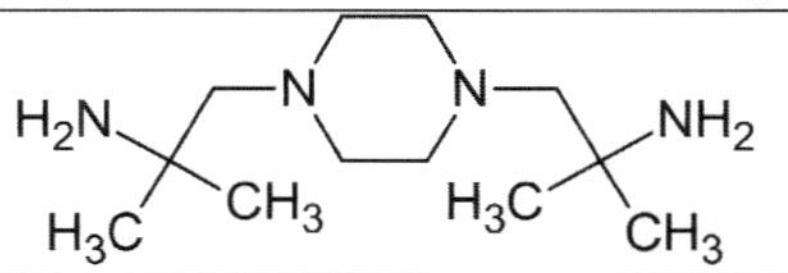
Monopropanolamine	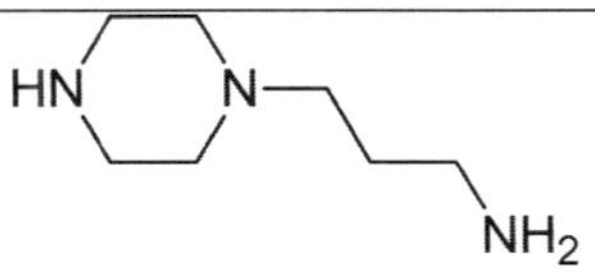	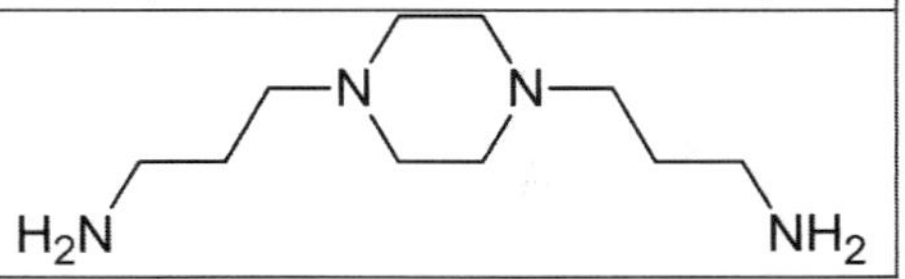

AEP has been studied as a potential capture solvent by Chen (Chen, 2011) and Du (Rochelle, 2012a; Rochelle, 2012b). 6 m AEP and 5 m PZ/2.3 m AEP both have lower cyclic CO_2 capacities than 8 m PZ but have competitive CO_2 absorption rates, slightly higher heats of absorption and, in the case of 5 m PZ/2.3 m AEP, have comparable thermal degradation characteristics as 8 m PZ. AEP, when blended with PZ, has better solid solubility characteristics than 8 m PZ, a key limitation of the concentrated PZ solvent. AEP (and other triamines and tetramines) are less volatile than PZ and thus will have reduced amine emissions from the CO_2 absorption column and/or the water wash column. Although AEP is commercially available, the other synthesis products, which are shown in in table 1.2, are not readily available. These synthesis products, however, can be produced in the regeneration section of the CO_2 capture plant.

2. Experimental Methods

Samples were prepared gravimetrically. CO_2 was sparged into the solution and measured gravimetrically. An initial CO_2 loading of 0.3 mol CO_2/mol alkalinity was used for each solution that was prepared. 7 ml of solution was loaded into 10 ml stainless-steel Swagelok® cylinders, which were then sealed and placed in convection ovens at set temperatures of 135 and 150 degrees C. These temperatures correspond to those encountered in the regeneration sections of proposed CO_2 capture plants using amine solvents. Cylinders were removed periodically throughout the experimental run.

A Dionex ICS-2100 chromatography unit was used to analyze for parent amine concentrations and synthesis product concentrations. Samples were diluted 10000X by mass. A Dionex CG17 50 mm guard column and a CS17 250 mm analytical column connected in series were used to carry out the separation 18.2 umho deionized water spiked with a gradient of methylsulfonicacid was used as the mobile phase. A Metrohm potentiometric autotitrator was used to estimate the loss of alkalinity in each sample. Samples were diluted 300X by mass and 0.2N H_2SO_4 was added until the pH reading reached approximately 3.9, corresponding to total protonation of all amine compounds present in the sample. The experimental methods are essentially similar to those of Freeman (Freeman, 2011).

A list of chemical suppliers and product purity is presented in table 2.1.

Table 2.1. List of chemicals suppliers and product purity

Chemical name	Purity	Supplier
Piperazine (PZ)	99	Sigma-Aldrich
Monoethanolamine (MEA)	99	Acros Organics
Monopropanolamine (MPA)	99	Acros Organics
Monoisopropanolamine (MIPA)	93	Sigma-Aldrich
2-Aminomethylpropanol (AMP)	99	Acros Organics
Methylaminoethanol (MAE)	99	Acros Organics

The supplied MIPA likely contains an isomer as its principal impurity (Bedell, personal communication, May 2012). This isomer is shown in figure 2.1 and can also participate in the amine synthesis like MIPA.

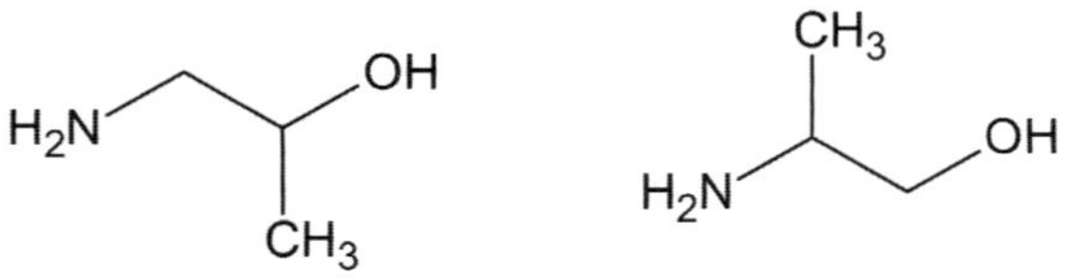

Figure 2.1. Isomers of MIPA; the principal product is on the left

3. Results and Analysis

Unified Calibration Curve

Standards were unavailable for several of the synthesis compounds made. The calibration data, however, for AEP and PZ can be modeled well using a single calibration curve. These calibration data were used to estimate the quantities of the synthesis products present in each sample; the curve is presented in figure 3.1.

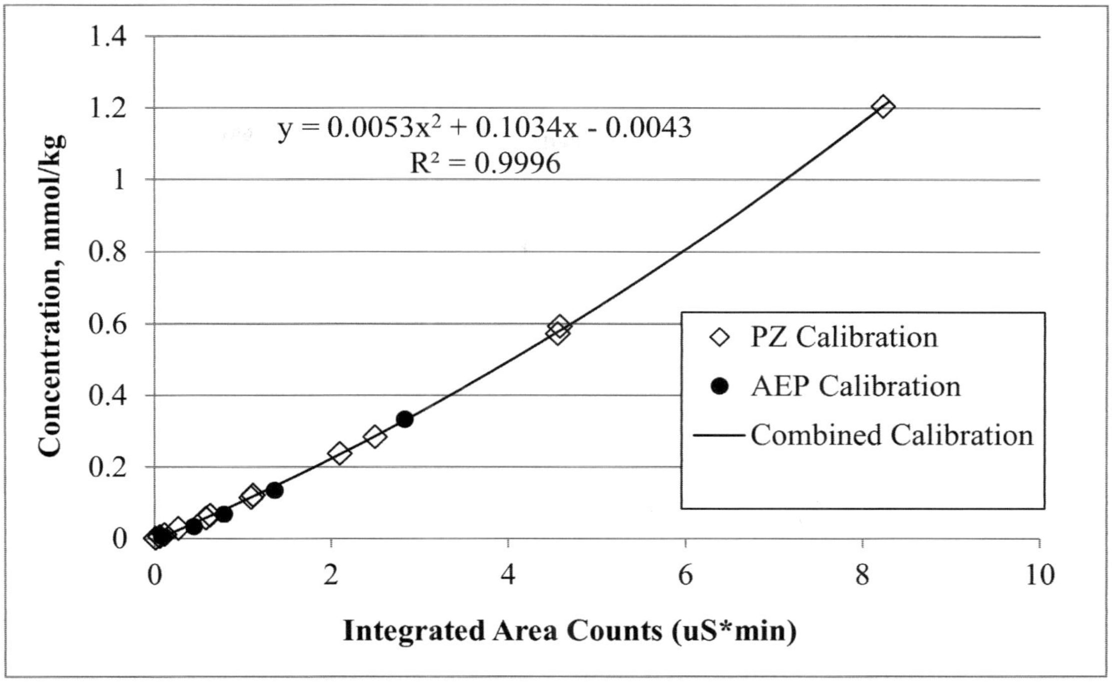

Figure 3.1. Unified Calibration Curve for PZ and AEP

Representative Chromatograms

A sample chromatogram of the synthesis reaction between 7 m PZ/2 m AMP initially at 0.3 mol CO_2/mol alkalinity after 1030 hours and 150°C is shown in figure 3.2. The parent amines and the synthesis product have been highlighted in the chromatogram.

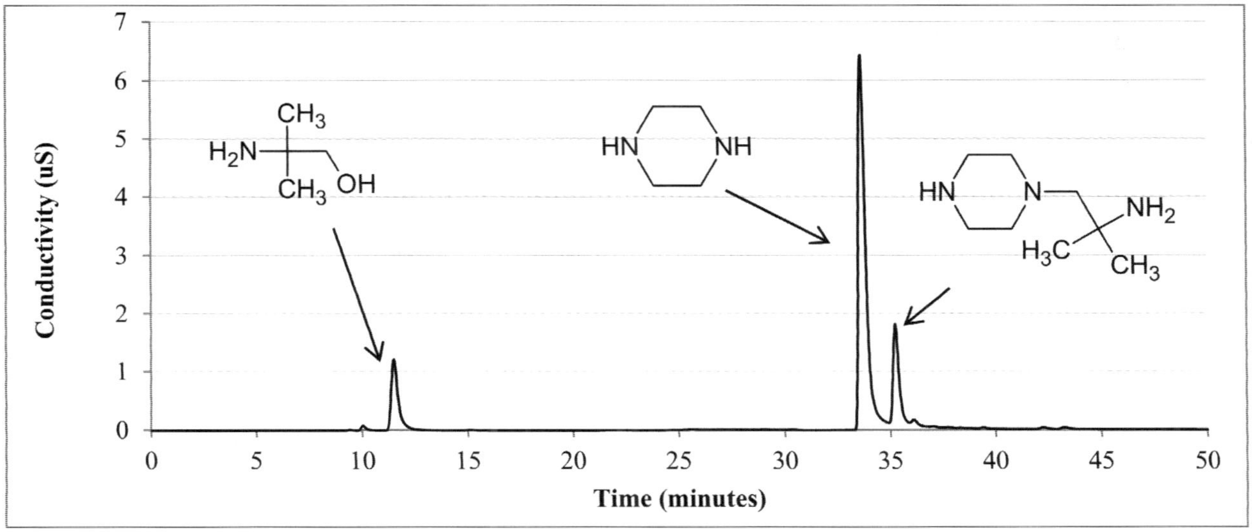

Figure 3.2. Sample chromatogram of AMP-PZ synthesis

Representative Reaction Behavior – First Order Modeling & Arrhenius Behavior

A first order rate model was used to estimate the loss of the oxazolidinone-forming amine throughout the reaction run. Figures 3.3 and 3.4 suggest that this model approximates amine loss reasonably well.

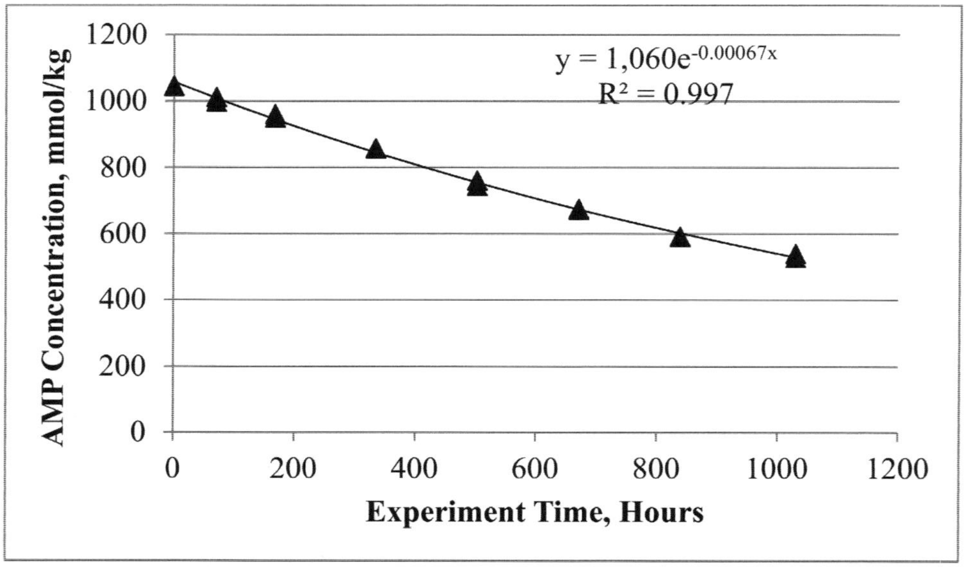

Figure 3.3. Concentration vs time plot of AMP in 7 m PZ/2 m AMP at 150°C, initially at 0.3 mol CO_2/mol alkalinity loading

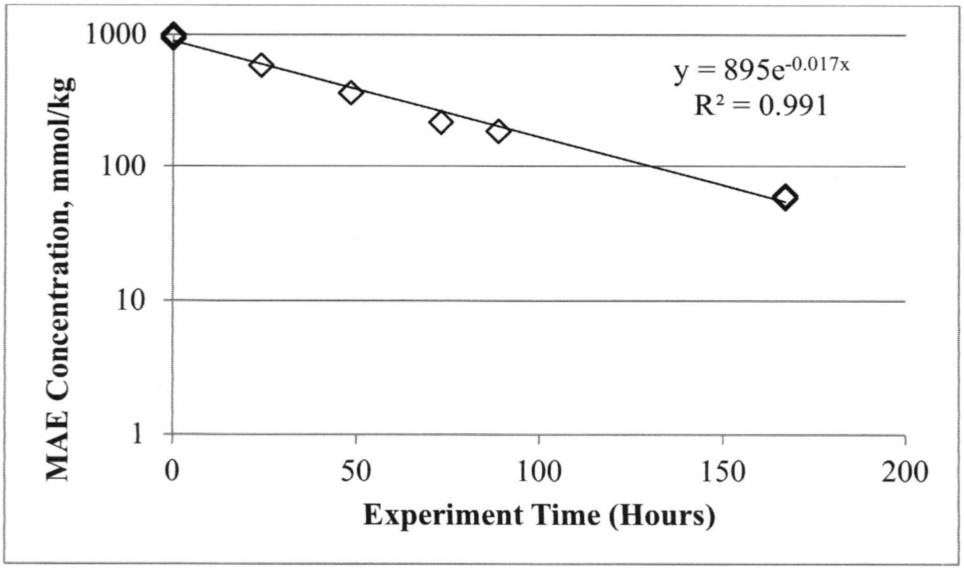

Figure 3.4. Concentration vs time plot of MAE in 7 m PZ/2 m MAE at 150°C, initially at 0.3 mol CO_2/mol alkalinity loading

The activation energy was estimated by using the Arrhenius relationship. Although this work only incorporates data from two different temperatures, data from Freeman suggests that thermal degradation of amines (among them oxazolidinone-formers) follow this behavior (Freeman, 2011).

Selectivity to the synthesis products is defined as the following:

$$\text{Selectivity}_t = \frac{\sum C_{\text{Synthesis Product},t}}{C_{\text{Oxazoidinone Former},i} - C_{\text{Oxazolidinone Former},t}} \quad (3.1)$$

Synthesis products can constitute both the triamine synthesis product and the tetramine dimer synthesis product shown in table 1.1.

Results Summary

Table 3.1 shows the summary of the results for the various amines tested.

Table 3.1 – Results Summary
Note: Degradation data apply to 150°C

Oxazolidinone-Forming Amine	MEA	MPA	MIPA	AMP	MAE
Amine Conversion (at 500 hours)	85%	45%	44%	28%	100% *(occurred between 167 and 340 hours)*
Synthesis Product Selectivity (at 500 hours)	69%	>90%	87%^	93%	>90%
Alkalinity Loss, %/week	2.0	1.7	1.5	1.0	5.0 (at 167 hours)*
First-order Amine Conversion Rate, 1/hr	0.0042	0.0012	0.00088	0.00067	0.017
Activation Energy, kJ/mol	160	150	130	140	100

* alkalinity of the solution was essentially unchanged after 167 hours (decreased by 0.2 %/wk).
^ also includes synthesis products from the secondary isomer

Analysis

Amine structure appears to have a strong effect on the rate of reaction with PZ. Increasing the steric hindrance of the oxazolidinone forming amine – either by lengthening the carbon chain from two to three carbons or by adding functional groups to the carbons in the alpha position relative to the amino or hydroxyl groups – reduces the rate of reaction with PZ by a factor of three to six. On the other hand, it appears as if oxazolidinone-forming secondary amines have a rate of reaction with PZ that is about four times as fast than the primary amines.

Alkalinity loss can be correlated with the oxazolidinone-forming amine conversion rate. Increased rates of alkalinity loss are indicative of the formation of ureas. Ureas can be formed from the synthesis product that is produced when an oxazolidinone-forming amine reacts with its own oxazolidinone. This is commonly encountered in MEA thermal degradation when the MEA-oxazolidinone synthesis product, hydroxyethylethylenediamine (HEEDA), can rapidly form a urea.

A key finding is that it that the alkalinity loss rate is reduced significantly after consumption of the oxazolidinone-forming amine – especially in the case of the PZ-MAE synthesis reaction. Once the MAE was consumed, the PZ loss was reduced to approximately 30 mmol/week and alkalinity loss was reduced by a factor of 25. This implies that the amine synthesis products and piperazine form a thermally stable mixture and therefore can be used at elevated temperatures in the regeneration section without significant amine loss. A similar phenomenon was noted in PZ-MEA synthesis, with PZ loss averaging about 10 mmol/week once the MEA was consumed.

4. Conclusions

a. First-order kinetics can accurately model the consumption (loss) of the oxazolidinone-forming amine when reacted with piperazine to form substituted triamine and tetramine piperazine analogues.
b. At 150°C, an initial amine concentration of 7 m piperazine and 2 m oxazolidinone forming amine, and at an initial loading of 0.3 mol CO_2/mol alkalinity, the oxazolidinone-forming amine consumption rate is the following:
   - Monoethanolamine: 0.0042 1/hr
   - Monopropanolamine: 0.0012 1/hr
   - Monoisopropanolamine: 0.00087 1/hr
   - 2-amino-2-methylpropanol: 0.00067 1/hr
   - Methylaminoethanol: 0.017 1/hr
c. The structure of the oxazolidinone-forming amine plays a significant role in its loss rate:
   - Increasing steric hindrance – such as increasing the chain length between the amino group or hydroxyl group or by adding alkyl functional groups to the carbon chain between the hydroxyl and amino groups – can reduce reaction rates by a factor of three to five.
   - Secondary amines that form oxazolidinones react four times as fast as primary oxazolidinone-forming amines.
d. Piperazine and solvent alkalinity losses are significantly reduced (~0.2 wt%/wk) once the oxazolidinone-forming amine is completely consumed, indicating thermal stability of the synthesized amine blend.

5. Acknowledgements

The authors wish to acknowledge the Luminant Carbon Management Program for funding.

6. References

Chen X. *Carbon Dioxide Thermodynamics, Kinetics, and Mass Transfer in Piperazine Derivatives and Other Amines*. The University of Texas at Austin. Ph.D. Dissertation. 2011.

Davis JD. *Thermal Degradation of Aqueous Amines Used for Carbon Dioxide Capture*. The University of Texas at Austin. Ph.D. Dissertation. 2009.

Freeman SA. *Thermal Degradation and Oxidation of Aqueous Piperazine for Carbon Dioxide Capture*. The University of Texas at Austin. Ph.D. Dissertation. 2011.

Gouedard C, Picq D, Launay F, Carrette PL. "Amine Degradation in CO2 Capture. I. A Review." *International Journal of Greenhouse Gas Control*. 2012;10:244-270.

Rochelle GT et al. "CO_2 Capture by Aqueous Absorption, Fourth Quarterly Progress Report 2011." Luminant Carbon Management Program. The University of Texas at Austin. 2012.

Rochelle GT et al. "CO_2 Capture by Aqueous Absorption, First Quarterly Progress Report 2012." Luminant Carbon Management Program. The University of Texas at Austin. 2012.

(95a) Accounting, Finance, and Statistics II

Makarand (Raj) Joshi, Organizational Strategy Management and Operations, Merck & Co., Inc., Whitehouse Station, NJ

This is a tutorial-like presentation of the MBA in a Day session. Please attend both talks, Accounting, Finance, and Statistics I and II, if you want to learn about all three subjects, which will be divided across the two talks. All three subjects will not be covered in each of the two talks.

The speaker is Makarand (Raj) Joshi, who is currently Lean Six Sigma Lead for Global Science, Technology, and Commercialization organization at Merck & Co. in New Jersey. He has extensive experience in R&D and Manufacture of Pharmaceuticals and Medical Devices and has led entire gamut of activities from Concept Identification through Product Development to Launch and Supply. Raj has a PhD in Chemical Engineering from Cornell University and started his career in research. Subsequently he moved to development and then to manufacturing. Mid way in his career he obtained an MBA degree from Lehigh University. He is a certified Black Belt and is currently pursuing a Master Black Belt certification.

The session will be of interest to those who want to learn about MBA programs and content: Those who may be considering enrolling in an MBA program, those who want to understand the type of information that is covered in an MBA program, and those who would benefit from having an overview of the content of a typical MBA program, for example. The speakers are subject matter experts who will take into account the interests and background of chemical engineers in the way that they present the information. The average student who is enrolled in an MBA program has six years of experience and is 29 years old. Students with engineering or science backgrounds comprise 35 to 40% of MBA programs. As a consequence, the session will be highly of interest to young professionals.

Effective Communication Based Leadership Skills in a Dynamic Global Market.

With globalization the world market is expanding and barriers between countries are dissolving. As a result, the demand of the world market is no longer influenced by market players of developed countries alone. The market players from developing and under-developed countries are becoming more influential. Doing business in one's domestic front alone is no longer a sound sustainable business strategy.

Two distinct yet complementary forces are shaping the world market: technology - innovation and market expansion across the globe. The convergence of these two forces is challenging global companies to: 1) offer higher quality, more reliable and more standardized products and services at a lower cost to thereby improve profitability; and 2) establish more market share. Innovations and product development extend world-wide and are expanding rapidly. Technological advances and competitive market pressures are making new product life-cycles shorter than ever before. The rate of assimilation of these advancements is decreasing, and product development costs are skyrocketing. Thus business dynamics are getting more complex and challenging than ever before. As this metamorphosis is taking shape and continuing, it is forcing the past equilibrium of these dynamic forces towards a new one.

With such transformations taking place, in order to manage these dynamics the leadership style itself is undergoing substantial changes. Such changes include a better understanding of cultural differences, business practices, business ethics and styles of the multicultural business players. Pro-active planning is becoming of paramount importance in understanding such transformations. Such understanding helps to identify necessary action steps and their subsequent implementation. It is becoming necessary to fine-tune the organizational structure to effectively achieve global presence with a local touch.

Among all the important traits of a successful leader, the most invaluable one is his/her skill at effective communication. Coupled with the necessary technical and/or business knowledge and understanding, effective communication skills are essential for business leaders to face this ever-changing global marketplace and to establish sustainable business success. Figure 1 below schematically describes a successful leadership style

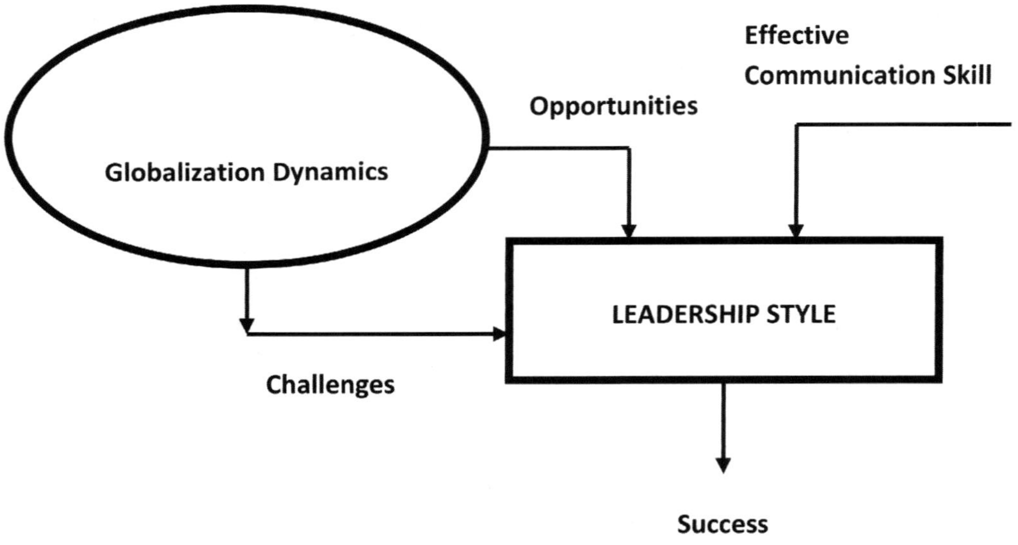
Effective
Communication Skill
Opportunities
Globalization Dynamics
LEADERSHIP STYLE
Challenges
Success

Operations, Supply Chain and Risk Management
Speaker: Deborah L. Grubbe, PE, CEng.

The competitiveness and health of a manufacturing based enterprise is critical to its sustainability and financial success. In this workshop, participants will learn effective principles for strong and sustainable operations, and will gain insight into the key areas of enterprise risk management. The workshop will focus on the following strategic subjects:

1. Creating and managing a sustainable operation – world class manufacturing
2. Concepts of Sustainability – applied to the enterprise
3. Values based thinking – safety and environmental leadership
4. Understanding supply chain principles and logistics
5. Making your workforce your competitive advantage

(95b) Strategy and Economics

Diana Matonis, Midstream Energy Group, Sugar Land, TX

This is a tutorial-like presentation of the MBA in a Day session. The speaker is Diana Matonis, PhD, MBA. She is Senior Engineer/Plant (Asset) Engineer at DCP Midstream. Dr. Matonis has more than 20 years of experience supporting the Oil & Gas industries. She has held key positions in both project and project management roles. She has developed techno-economic and life cycle assessments and has provided technical strategies and solutions to bring to market licensed patents through government and industrial partnerships. Technical areas of expertise include new, lower-cost NGL and syngas processes, conversion of biomass feedstock, techno-economic and life cycle assessment of bioenergy systems, and process performance models for scale-up. Besides DCP Midstream, Dr. Matonis has held management and technical positions with Aerotek Engineering, Fuel Cell Energy at the Aries Group, the Gas Technology Institute, and Argonne National Laboratory.

The goal of this presentation will be to provide a basic introduction of key macroeconomic variables involved in assessing the macroeconomic environment in which a business operates. Macroeconomic conditions play important roles in business decisions and performance. Aggregate income, unemployment, and inflation rate influence profitability. Interest rates determine the cost of capital, and exchange rates affect international competitiveness.

Thus, these key macroeconomic variables will be covered;

♦ Gross Domestic Product (GDP)

♦ Unemployment Rate

♦ Inflation Rate

♦ Interest Rate

♦ Exchange Rate

The session will be of interest to those who want to learn about MBA programs and content. The session will be highly of interest to young professionals.

(120a) Hybrid Cancer Therapeutics

Deniz Cetin and **Andrew Pike**, Department of Chemical Engineering, Lehigh University, Bethlehem, PA

Andrew Pike

Deniz Cetin

Project: Hybrid Cancer Therapeutics

Lehigh University

Department of Chemical Engineering

Advisors: Professors Bryan W. Berger & Mark A. Snyder

Abstract:

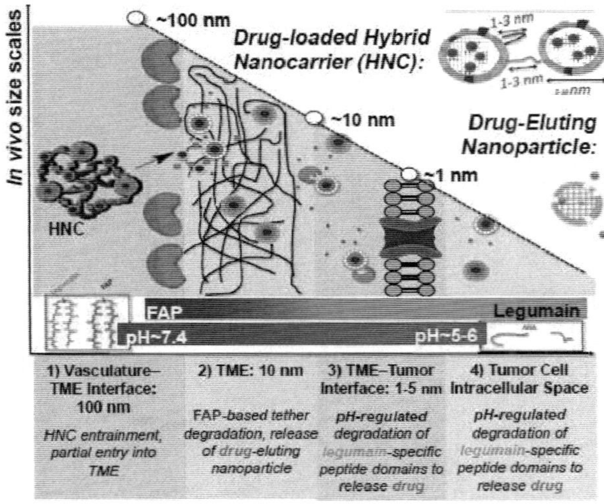

In nano-scale drug delivery, when targeting cancer cells, a recurring challenge is that the more drug-resistant cancer cells survive the initial burst of drug and continue to metastasize into more resilient tumors. There are several size selective interfaces along the delivery pathway that the delivery system will be designed to permeate. For this reason, a delivery system with multiple phases of delivery is desirable. A solution to this is to control the rate of drug delivery through particle size, porosity, and architecture. In addition, pH responsive proteins can be used to target tumor sites and prevent drug release to healthy cell environments. Current approaches include organic particles tethered in order to target specific ligand, which create problems such as chemical instability, inefficiency in use of drug (reduced half-life *in vivo*), eliciting an immune response, and variable porosity. Our project focuses on tethering of mesoporous silica nanoparticles with peptides which change conformation when exposed to environments of different pH values and have certain protease-cleavage sites, with the purpose of creating an initial burst of cancer drug-carrier nanoparticles with sustained release.

Our approach seeks to minimize these problems by using particles of 5 nm in diameter to control the release of drug, while designing the peptide structure to react to changes in the tumor microenvironment. This ensures that the silica nanoparticles are not exposed to the environment and are not releasing drug until the delivery system reaches a tumor environment.

(120b) Improving the Mechanical Properties of Activated Carbon Nanofiber Nonwovens

Breanne Muratori, Department of Chemical, Materials, and Biomolecular Engineering, University Of Connecticut, Storrs, CT, Seetha S. Manickam, Chemical, Materials, and Biomolecular Engineering, University Of Connecticut, Storrs, CT and Jeffrey R. McCutcheon, Department of Chemical, Materials and Biomolecular Engineering, University of Connecticut, Storrs, CT

Activated Carbon Nanofiber Nonwovens (ACNFNs) have a wide range of potential applications because of their unique properties like high specific surface area and good electrical conductivity. Potential applications include use as adsorption media, electrodes in microbial fuel cells, tissue scaffolds, and many others. A critical aspect that hinders ACNFNs from being used more widely in these applications is their weak mechanical strength. The goal of this work is to improve the strength of ACNFNs, without sacrificing surface area and conductivity.

Several different methods of improving strength were explored. The methods of fabrication were examined, looking for areas where the process could be altered, yielding a material with greater strength. The fabrication of ACNFNs is a four-step process: electrospinning, stabilization, carbonization, and steam activation. Alternative polymers for electrospinning, effects of hot-stretching during pyrolysis, and varying carbonization and activation protocol were studied. Samples were tested for mechanical strength, carbon yield, and surface area.

(120c) Nano-Confined CO2 Sorbents for High-Efficiency CO2-Capture

David Palm, Chemical Engineering Department, University of Pittsburgh, Pittsburgh, PA, Karen J. Uffalussy, Chemical and Petroleum Engineering, University of Pittsburgh, Pittsburgh, PA, Robert M. Enick, Department of Chemical and petroleum Engineering, University of Pittsburgh, Pittsburgh, PA; Chemical and Petroleum Engineering Department, University of Pittsburgh, Pittsburgh, PA and Götz Veser, Department of Chemical and Petroleum Engineering, University of Pittsburgh, Pittsburgh, PA

CO_2 capture and sequestration are among the prime challenges of our time, as the need to reduce atmospheric concentrations of this greenhouse gas has become evident. Among the most mature CO_2 capture technologies are liquid amine-based processes; however, even these applications are hampered by high cost and large energy penalty. Grafting or supporting liquid amines on solid supports offers a promising avenue to overcome some of these limitations. In particular, embedding amines into microporous solids constitutes a simple, flexible approach.

In the present project, we are investigating the application of this embedding approach to a novel class of CO_2 sorbents, so-called phase-change sorbent materials. An aminosilicone sorbent ("GAP-0") that has recently shown great promise as a CO_2 sorbent due to its high sorbent capacity and its ability to change from a liquid to a solid upon CO_2 sorption, was chosen as a model compound and embedded into porous silica shells. The resulting material has the advantage of a well-defined particle size, suppressed agglomeration, and fast CO_2 transport. Furthermore, on a fundamental level, the impact of nanoconfinement on phase transitions of materials is poorly understood to-date and the present system poses a well-defined model system for such studies.

GAP-0 was loaded into silica "nanobubbles" with ~30 nm diameter via a straight-forward wet impregnation method. The walls of these nanobubbles are ~10 nm thick and highly porous with average pore diameters of ~8Å. After characterization of the material, the CO_2 adsorption capability of the sorbent was studied via Thermo-Gravimetric Analysis (TGA) during cyclic CO_2 uptake and release over a temperature range between 45-90°C.

Results to-date show that nanoencapsulated GAP-0 does indeed adsorb and desorb CO_2 effectively. However, the GAP-0 compound exhibited unexpected volatility over the temperature range of the experiments, leading to a loss of sorbent material over time. These losses were reduced via optimization of uptake and release conditions (time, temperature), but multi-cycle experiments still showed continual loss of CO_2 sorption capacity. Current work is extending the demonstrated approach onto other aminosilicone-based phase-change sorbent materials.

(120d) Using Porous Carbon Nanotube Membranes for Separation of CH4/CO2 and CH4/H2 Mixtures

Benjamin Bucior[1], Jinchen Liu[2], De-Li Chen[1], De-en Jiang[3] and J. Karl Johnson[1], (1)Dept. of Chemical and Petroleum Engineering, University of Pittsburgh, Pittsburgh, PA, (2)University of Pittsburgh, Pittsburgh, PA, (3)Chemical Sciences Division, Oak Ridge National Laboratory, Oak Ridge, TN

Although separations are typically expensive and energy-intensive, membranes have the potential to separate gases both economically and efficiently. However, conventional polymeric membranes exhibit a tradeoff between selectivity and permeance. Carbon nanotubes (CNTs) exhibit very different transport behavior compared with other nanoporous materials, having transport diffusivities orders of magnitude higher than other matierals. Theoretically, gas mixtures could be separated by pores in a CNT membrane via a size-sieving mechanism. This means that nanoporous CNT membranes have the potential to be both selective and permeable, overcoming the limitations of conventional membranes.

We modeled a theoretical porous carbon nanotube (PCNT) membrane by generating pores on the wall of a CNT. We used molecular dynamics simulations to study fluid transport across the PCNT membrane, comparing models with rigid and flexible bonds. We also calculated the potential energy surfaces of each fluid molecule passing through the PCNT. Our simulations of both rigid and flexible PCNTs showed high selectivity for CH4/H2 and CH4/CO2 mixtures. Separations of N2/CO2 mixtures were also studied.

(120e) Determination of Degree of Polymerization of Cellulose Using MALDI-TOF with a Novel Ionic Liquid Matrix

Michael Mayer, University of Toledo, Toledo, OH, B. Leif Hanson, Instrumentation Center, The University of Toledo, Toledo, OH, Wendell Griffith, University of Toledo, Toledo and Constance Schall, Chemical & Environmental Engineering, University of Toledo, Toledo, OH

Polysaccharides, such as cellulose, have been investigated as a source for production of renewable fuels and other carbon based materials. Characterizing cellulose, especially its degree of polymerization (DP), is an important step in evaluating its deconstruction to monomeric sugars or in its use for materials. Matrix Assisted Laser Desorption Ionization Time of Flight Mass Spectroscopy (MALDI-TOF) has been used to determine the degree of polymerization of many polymersandmore recently carbohydrates. The use of MALDI-TOF for cellulosic analytes has been difficult due to their insolubility in traditional matrix solutions. A new technique has been developed for MALDI-TOF, utilizing an ionic liquid matrix and solvent system. This approach was used to determine DP in an excess of 200 for crystalline cellulose. The technique has potentially wide application in analysis of lignocellulosic substrates.

Abstract

Global air pollution is a pandemic phenomenon that threatens both the stability of local ecosystems and the longevity of the human race. Carbon dioxide is the main pollutant of our atmosphere and is produced by both living beings and the burning of fossil fuels; the abundance of carbon dioxide in the atmosphere has resulted in temperatures and sea levels achieving an all time high. A solution proposed is that of creating a carbon dioxide consuming biomaterial. For this study, chloroplasts from spinach leaves were isolated and incorporated into silk matrices to create such a material. Both silk-chloroplast hybrid films and hydrogels were examined using microscopic analysis to determine the optimal environment for chloroplast survival outside of the plant cell. Hydrogels were chosen as the primary medium through which to incorporate chloroplasts into silk matrices as it provided an aqueous biomimetic environment that optimized chloroplast survival and functionality outside of the plant cells.

From there, tests were performed to determine the functionality of both hybrid films and hydrogels outside of the cell. These tests used apparatuses through which cigarette smoke was passed, the carbon dioxide filtering ability of the silk-chloroplast hybrid being a marker of the system functionality. The hybrid films were tested using a less-advanced apparatus than the hydrogels, as it was expected that they would not be as functional. For this test, cigarette smoke was passed through an apparatus and collected onto a pre-prepared silk mass. This silk mass was then processed and used to create films that were then analyzed using microscopy. From the microscopic analysis, it was apparent that the silk films were minimally functional as there was a marked decrease in particular deposits for the filtered group when compared to the control group. However, because of the non-aqueous environment of the silk-chloroplast hybrid film, it is unlikely that the silk-film hybrid would be functional for long-term applications.

The silk-chloroplast hydrogels were then tested using a more complicated apparatus. For this apparatus, the law of mass action was utilized by evacuating water tanks and drawing cigarette smoke into said water tanks, past the hydrogels. The pH of said water was then tested against a control and against pure water to determine the functionality of the silk-chloroplast hybrid hydrogels. It was expected that carbonic acid would form in the water tanks, as it is a product produced by the reaction of carbon dioxide and water. Therefore, the filtered water tanks were expected to have less carbon dioxide reacted with water, thereby having a higher pH. This result was tested using litmus pH strips and the expected result was achieved.

From this research, future investigation of the filtering capabilities of silk-chloroplast hybrid films can be performed to address the growing global concern of atmospheric pollution. Genetic recombination of chloroplasts and the sophistication of hydrogel functionality and test apparatuses are possible avenues of further research for the continuation of this project.

(169a) From Cancer to Hydrogen: Cell-free Biomolecular and Bioprocess Engineering

James R. Swartz, Department of Chemical Engineering, Stanford University, Stanford, CA

Normally, the design and production of cancer therapeutics would seem to have little in common with large-scale hydrogen production. Yet, the amazing capabilities of biological systems provide the components and principles for both pursuits. In particular, cell-free methods allow us to manipulate and harness even very complex biological reaction systems. In this presentation, I will describe progress toward: a) new cancer diagnostics and novel cancer vaccines, b) a potent and broadly protective influenza vaccine, and c) a potential highly productive process for sustainable hydrogen production from cellulosic biomass. Even though the applications are highly diverse, they can all be addressed by the power of fundamental biochemical engineering.

(169c) Biochemical Engineering in Half a Century: Tracing the Steps of Prof. Daniel IC Wang

Wei-Shou Hu, Chemical Engineering and Materials Science, University of Minnesota, Minneapolis, MN

For over five decades Prof. Wang has devoted himself to biochemical engineering education, research, and technology creation. The course of his career illustrates the tremendous advances that have taken place in the field. His work helped propel technological innovation and development as it reached critical junctures. This presentation will use Prof. DIC Wang's work to highlight the advances in biochemical engineering in the past half a century and invite the audience to reflect on its impact on the future of the profession.

(205a) Rare Earth Elements: A US User's Perspective of Issues and Initiatives

Peter Dent, Electron Energy Corporation, Landisville, PA

Strategic and critical mineral supply chains are nationally recognized as being important to the US manufacturing industry, energy supply, and national defense. Currently, China dominates the rare earth market and will continue to do so even after new non-Chinese sources of supply come on line. The17 rare earth elements play a crucial role in international supply chains. The largest US industrial consumers by tonnage are catalyst producers who use rare earths for the cracking of crude petroleum and automotive catalytic converters for treating engine exhaust. Growing demand for alternative energy technologies which utilize rare earths include wind power, hybrid electric vehicles, advanced batteries, fluorescent lighting, and efficient permanent magnet motors and generators. To secure the future of these industries, we must address the challenges associated with supply risks due to their location, vulnerability to disruptions, and lack of suitable substitutes. This presentation will discuss the issues and initiatives in supply and demand in the complete rare earth supply chain and offer a user's perspective on navigating through this uncertain environment.

(205b) The Role of a Transformed National Defense Stockpile in U.S. Strategic Materials Security

Richard A. Lowden, Materials Science and Technology Division, Oak Ridge National Laboratory, Oak Ridge, TN

The United States established a stockpile of strategic materials in 1939 beginning with a handful of ores and minerals which with time was expanded to include additional minerals, metals, chemicals, and other processed materials. The purpose of the stockpile was to preclude a dangerous and costly dependence by the United States upon foreign sources for supplies of scarce but important materials in times of national emergency. Although emphasis has traditionally been on the acquisition and retention of certain strategic and critical materials, the legislation includes provisions that encourage the conservation and development of resources and materials. The authors of the Stockpiling Act realized the value of recycling, improved separation techniques, and alternative methods for refining or processing of materials thus included language to address these issues. Since 1993 the Stockpile has been downsizing, selling off most of its assets, however, recent changes in the global marketplace and increasing risks associated with the supply of important materials such as the rare earths, interest in the stockpile has been renewed. The Stockpile is in the process of being transformed in a more flexible and dynamic strategic materials risk management (security) program with much broader scope and applicability with respect to the entire supply chains for important materials.

(205c) Material Sustainability At General Electric

Anthony Ku and Steven Duclos, GE Global Research, Niskayuna, NY

GE Global Research, working with our GE business partners, has developed a comprehensive strategy to mitigate risks in the availability and sustainability of raw materials that are critical to high technology products. Many such products contain rare elements with limited availability due either to geopolitical constraints or the fact that they are refined as by-products of limited growth commodity metals. We have developed a methodology to assess such risks and prioritize technical solutions. These solutions include the development of advanced manufacturing techniques that minimize waste, recycling technologies to recover materials lost during manufacture or at end of life, and materials engineering that enables substitutions resulting in significant reduction in the use of at-risk elements without sacrificing material properties. Recent examples include our development of superalloys with 50% and 100% less rhenium, and magnet materials with reduced rare earth content. This presentation will discuss GE Global Research's work in this area.

(205d) The Issues and Opportunities Associated with Recycle and Reuse of Strategic and Critical Materials

Eric S. Peterson, Chemical Sciences, Idaho National Laboratory, Idaho Falls, ID

Strategic and critical mineral supply chains are nationally recognized as being important to the US manufacturing industry, energy supply, and national defense. For example, the largest industrial consumers of rare earths in the U.S. are catalyst manufacturers, representing almost 30% of the domestic market by volume in 2008 (40 to 45% in the previous five years). Rare earth elements are also essential to the production of catalysts for the cracking of crude petroleum and of automotive catalytic converters for treating engine exhaust and catalyst that contain rare earth elements influence the production of a diverse range of end user products from aircraft windows to dentures, plastic food packaging, etc. Many facets of the strategic and critical materials supply chain involve "chemistry" or "chemical engineering" from novel digestion and separation processes to reduction and refining to utilization in "old" and "new" products and technologies as well as recycle and reuse. This presentation discusses the issues and opportunities associated with recycle and reuse of critical materials.

(205e) Roundtable Discussion On Critical Materials Supply Chain and Sustainability

Sharon Robinson, Oak Ridge National Laboratory, Oak Ridge, TN

Strategic and critical mineral supply chains are nationally recognized as being important to the US manufacturing industry, energy supply, and national defense. Many facets of the strategic and critical materials supply chain involve chemistry and chemical engineering from novel digestion and separation processes to reduction and refining to utilization in "old" and "new" products and technologies as well as recycle and reuse. After a series of papers from leading experts in the field from both government and industry, the presenters will hold a round-table discussion to identify the issues and opportunities associated with the critical materials supply and sustainability.

(252a) Clicking Polymers Together: Assembly of Complex, Controlled Polymer Structures from Efficient Chemistries

Christopher N. Bowman, Chemical and Biological Engineering, University of Colorado, Boulder, CO

A new paradigm encompassing several distinct chemical reactions and, more importantly, a generalized approach to molecular design and synthesis has been rapidly adopted in the fields of chemical synthesis, biotechnology, materials science, drug discovery, surface science, and polymer synthesis and modification. The *Click Chemistry* paradigm focuses on implementation of highly efficient reactions that achieve quantitative conversion under mild conditions. As such, these reactions represent ideal candidates for further development, understanding and implementation. In particular, the synergistic combination of these click chemistries with photochemical initiation and polymer formation has been used to afford 4D control of polymer formation, structure and patterned assembly. Here, we will focus on three distinct vignettes related to our implementation of photoclickable polymer systems. The first of these focuses on the development of covalent adaptable networks (CANs) where the ability to controllably alter the network structure is used to alter topography and other material properties by forming materials which can be switched revesibly from elastic to plastic simply by exposure to light. Secondly, we will focus on the development of approaches to photoinitiate the Cu(I) catalyzed azide-alkyne cycloaddition (CuAAC) click reaction. Here, implementation of this reaction in surface modification, hydrogel formation, and lithography as well as in the development of a new class of photopolymerization reactions will be presented. Finally, the development and implementation of click nucleic acids (CNAs) based on the thiol-ene click reaction will be presented. This distinct class of oligonucelotides combines the vast advantages of synthetic oligonucleotides such as peptide nucleic acids with the power of click reaction chemistry to form materials that hybridize with both natural and synthetic olignonucleoitides via Watson-Crick base pairing while being simple to produce in large scales appropriate for directed assembly and other high value materials applications.

(386a) Generating Complex Biological Traits through Combinatorial and Multigenome Expansion of the Sampling Space

Eleftherios T. Papoutsakis, Dept. of Chemical Engineering & Delaware Biotechnology Institute, University of Delaware, Newark, DE

There are many complex phenotypes/strains that are desirable for practical applications in the context of **Cell** or **Metabolic Engineering**. One example is when one desires to endow a platform organism with a biosynthetic or catabolic pathway that another organism may possess. Such a pathway may involve several enzyme-coding genes as well as regulatory genes, but these genes may be unknown. Another example is the ability to endow a platform organism with unusual membrane structures or a stress-response system that some other organism possess, or by re-engineering a native stress response system. This would be important in bioprocessing and advanced bioremediation applications. Endowing cells with such desirable capabilities requires several or many genes deriving from other organisms, including genes from targeted metagenomic libraries. Whether sampling a single-genome space or a multi-genome space, a key difficulty that must be overcome is the ability to sample the combinatorial space of genetic-loci interactions that are necessary for developing the synthetic or semi-synthetic phenotype. How does one attack this class of problems in the context of synthetic biology and cell engineering?

(426a) Chemical Engineering in India: Challenges and Opportunities

Ganapati D. Yadav, Department of Chemical Engineering, Institute of Chemical Technology, Mumbai, India

Chemical Engineering has a very interesting history in India and has evolved from an applied chemistry perspective to a highly science based engineering profession, typically modelled on the ideas promoted by George Davies in England. When officially, the species of 'chemical engineer' did not exist in British industries until 1957 and the IChemE (UK) did not get charter from the government there, India had already well established chemical engineering programmes and jobs in industry. Several universities and now well known institutes were established and Chemical Engineering was much sought after degree programme over the year and lost its sheen after early 1990s due to the onslaught of 'computer' related courses. However, Chemical Engineering education being highly versatile and all-encompassing, its scope has increased in the context of current global challenges.

Production and supply of energy, without harming the environment, in tune with the demand is critical to sustainable development and poverty alleviation endeavours by India. Energy sector influences social, economic, geo-political and environmental aspects of development, including employment, accessibility to potable, agricultural and industrial water, agricultural productivity, food supply and preservation, transport, health, population profile, education, and industrial growth. India's burgeoning population, diverse culture and varied income groups are a very big challenge to make sustainable and equitable development. To provide for basic needs on one hand and 'entertain' the middle class, whose size in on the rising curve, will always be problematic since the material and energy needs will grow tremendously. In relation to sustainable development, including apprehensions of global warming, environmental effects and the problem of wastes, need to be addressed while trying to meet the energy aspirations. Safety in operation is also a matter of great concern even with proven technologies, as well as the broad and indefinite aspect of maximising the options available to future generations. Another aspect is the geographical location and availability of land in putting mega projects is a contagious issues which have several social and political ramifications. While continuing to use old technologies, the biggest new sector will be carbon management; in particular, carbon dioxide capture from fossil energy production facilities and its ultimate sequestration. Most likely, efficient carbon management will depend significantly on chemical processing technology and expertise.

The technologies related to producing advanced materials, clean energy generation and storage, medicines, high-end drugs, neutraceuticals, food products, fertilizers, agrochemicals, polymers, surface coating materials, laser dyes, colorants, pigments, adhesives, textiles, fibres, oleochemicals, surfactants, lubricants, water treatment and purification, air pollution abatement, bio-processing, downstream processing and a myriad of related issues involve high degree of science and engineering. How are we going to feed billions of people, remain in harmony with nature, and develop sustainable processes and technology? What will be their energy and material needs? Life expectancy is getting extended. Addressing these challenges requires a multifaceted effort that traverses the fields of chemistry, engineering, biotechnology, information technology and nanotechnology, engineering mathematics, environmental engineering and the curriculum and thus Chemical Engineering needs to widen its content without losing on the traditional areas.

(426b) Sustainability in the Chemical Industry through Collaborative Process Engineering

George Stephanopoulos, Massachusetts Institute of Technology, Cambridge, MA

Process Engineering has undergone significant transformation over the last 30 years, driven by chemical engineering innovations and collaborative engineering best practices, catalysed by software innovations. These transformations have created tremendous value for many companies resulting in capital and energy savings, together with increased safety, reliability and sustainability.

This presentation will discuss specific innovations and engineering collaboration best practices, highlight examples of recent successes in the industry and opportunities ahead. It will cover experiences across the asset lifecycle including R&D, conceptual engineering, basic engineering and engineering support to manufacturing and planning.

Chemical companies face an incredibly challenging and highly competitive environment. They need to explore ways to reduce costs associated with plant design, operation, maintenance and efficient use of energy, whilst increasing sustainability of their plant assets and their businesses. In this presentation we will highlight how they can take advantage of the latest innovations and collaborative engineering technologies to gain a competitive advantage in the global marketplace and the vision ahead.

(426c) Perspective of Solar Thermal Energy in India

Jyeshtharaj B. Joshi, University Institute of Chemical Technology, Mumbai, India, S. V. Panse, Institute of Chemical Technology, Mumbai, India and Anil Kakodkar, Bhabha Atomic Research Centre, Mumbai, India

India is facing an uphill task of improving standard of living of its 1.2 billion plus population. One major requirement for this purpose is to step up its per capita electric power consumption, which is 650-700 kWh at present. This is one fourth the world average, and fourteen times less than that for developed countries. To achieve a reasonable standard of living, India needs to upgrade its power generation capacity to provide for per capita electricity consumption of at least 5000kWh, which means electricity generation capacity of about 1.5TW. India today already stands fifth in the world in electricity generation, only preceded by China, U.S., Japan and Russia. The current global production of electricity is around 5 TW, and is expected to grow to around 9 TW by 2050.

India's main fossil fuel reserves are of cocking and non-cocking coal. Proven reserves are of 82 billion tons, and indicated reserves are 211 billion tons, only half of which is mineable. The proven oil reserves are of just 5.7 billion tons and indicated reserves are 30 billion tons. If India's power generation program heavily depends on fossil fuels, it can sustain, with the fuel consumption at the present rate of power generation, for a maximum period of two centuries, which itself is a short period by any standards in the life of a civilization 5000 years old. Further, it is alarming that the same period would reduce to barely 20 years, if only the mineable reserves and the increased fuel consumption for the proposed higher power generation is included in the picture. Thus, fossil fuel based power generation cannot be a sustainable solution for India (and, in fact, for the entire world). Also, unchecked consumption of fossil fuels would accelerate the process of global warming, and may bring us closer to the 'tipping point', after which the damage is irreversible.

Solar energy is a viable alternative for India's sustainable development, and environment friendly power generation technology. Indian subcontinent receives intense sunshine for many days in a year. The average DNI values in India range from around 6.4kWh/sq. m. to around 4.4kWh/ sq. m. depending upon location. It has been estimated that with 10% efficiency of conversion, solar energy incident on only 0.5% of land area can produce electricity at the total current capacity of 0.18TW.

Photo Voltaic panels convert solar energy directly into electricity, and need little water for operation. Because of continuous R&D, and economy of scale, its costs are continuously coming down. Today, one MW PV power plant costs $1.8 million, and power can be profitably sold at a price of $0.14/kWh. However, the alternative path of 'Concentrating Solar Power' (CSP) technologies has certain advantages of its own-namely easy and inexpensive energy storage, ability of generating process heat and facility of integrating with existing thermal power plants- and should also remain a significant component of India's solar energy program.

CSP technology was introduced in India, though on a miniscule scale as 'Scheffler Dish Technology'. It is a parabolic dish with one-way tracking, and a fixed receiver; and can generate steam up to about 120^0C. It has been installed in a few community kitchens in India for solar steam cooking. Domestic solar cookers were introduced in 1980s, and there was an aggressive extension program for them. But the program did not yield lasting and expected results. India did not have solar specific policies, until recent past. In the early period after its independence, apart from thermal power, the emphasis was on tapping the hydro-power potential, and developing technology for atomic power. Renewable energy sources in general and solar in particular, had not attracted attention of policy makers until last decade of twentieth century. In 1988 "Kutir Jyoti Scheme' was introduced by the Government of India (GOI), which aimed at extending electric connections to the households of rural poor families, with an emphasis on disadvantaged sections of society. 'Rajiv Gandhi Rural Electrification Scheme' introduced in 2005 was a major thrust towards rural electrification. In addition to grid extension, the programs of rural electrification included, on a demonstrative scale, PV based off grid solar electricity generation. In 2007, India announced its semiconductor policy to encourage semiconductor manufacturing, which also included solar PV panels.

But the real thrust of India in solar power has come in the form of 'Jawaharlal Nehru National Solar Mission'

(JNNSM). The Mission document was released on 14th Nov. 2009. It has an ambitious plan of generating 20,000MW solar power by 2020. The Mission period is divided into 3 phases: 2010-13 is phase one, 2013-17 is phase two, and 2017-22 is phase three. Unlike the earlier schemes of deploying solar technology only for off grid, rural electrification, the JNNSM also has ambitious targets for grid connected power generation via the solar route. In fact, grid power generation targets far exceed the off grid targets in the Mission. Thus, in phase one the target for solar grid power production is 1100MW, whereas that for off grid power production is 200MW. The same figures for phase two, and phase three are 10,000MW and 1,000MW; and 20,000MW, and 2000MW respectively. Also, the solar collectors (for electricity generation as well as process heat) are expected to cover 20, 125, and 240 million square meters of area respectively in the three phases.

In addition, GoI has recently (in May 2012) launched 'Commercialization of Solar Energy in Urban and Industrial Areas – ComSolar' initiative, in collaboration with the Government of Germany, and United Nations. Under the initiative, ten industries have been identified in India, which can preferably benefit from solar thermal technologies for supply of industrial process heat. It has been estimated that the solar potential in these industries is 600 kilo tons of oil equivalent (ktoe) per annum, and can result into a cost saving of nearly $300 million/annum.

It is too early to judge the outcome or impact of these initiatives. One observation, however, is that, in the process of reverse bidding during the first phase of JNNSM, the solar electricity cost in India has come down by nearly 50% of what was proposed while launching JNNSM. One difficulty experienced by the installers of solar plants in the first phase of the JNNSM is that, no reliable ground data of the DNI values is available in India. (The Indian Meteorological Department has now started collecting the data at 45 locations, spread across the country.) Also, unlike its successful program in atomic energy, the solar energy program of India had just become an exercise in policy framework and economic incentives; systematic R&D activity in the country is at its initial stage.

A major initiative is required to be taken towards development of large capacity solar thermal plants in India. Several India specific issues such as high dust load, sand abrasion, non-availability of water etc. would need to be addressed while also realizing competitive capital costs to enable large scale deployment of this technology. In addition to sustaining relevant research in a number of laboratories and industries; this would require setting up demonstration plants, bringing together combined capability of Indian R & D institutions and industry, on a large enough scale to permit commercial scale viability assessment. Creating such demonstration facilities alongside an academic institution with large focus on solar thermal technology, would realize the twin objective of human resource development and sustain further R & D on the basis of field data and experience. In particular, it has been proposed to establish three $5MW_e$ solar thermal power plants for testing and demonstration purpose. They will be based upon Power Tower (beam up and beam down) and Linear Fresnel Reflector technologies. It is proposed that these demonstration plants will be supported by a consortium of public/private, academic/research institutions that carry out relevant R & D in solar thermal technology and will start functioning within three years. The consortium members will in turn benefit from the experience and the performance data obtained on the demonstration plants. The centre will also develop standard testing protocols for CSP components. These initiatives will give a boost to R&D in the area of solar energy technologies in general, and CSP in particular.

For successful implementation of solar energy, the capital cost needs to be less than USD one million/MW. For solar thermal, R and D efforts are ongoing to improve the efficiency and the selection of cost effective materials. Research is focused on surface coatings (for reflectors) having the properties of good reflection, UV and erosion resistance and self cleaning. Research is also focused on surface coating for collectors, which can stand 800^0C and continues to have good absorbance and low emmitance. Other important areas of R&D are reduction in heat loss, PV-Solar Thermal Hybrids, low cost tracking, efficient land use, cost effective and efficient air condensers, heat transfer medium and the selection of thermodynamic cycle.

(426d) The Tiger and the Eagle: Partners in a Sustainable Energy Future

Jim Rekoske, UOP, Des Plaines, IL

In this talk we will compare and contrast the trajectory of energy sources and uses in India and the United States, illustrating that there is much to be learned from the past, and much to be shared in the future. Two case studies will be presented highlighting UOP's small contributions to this shared future, culminating in a discussion of our vision for how these cooperative efforts will lead to sustainable energy advantages for both the US and India.

(426e) Sustainability and Resilience of Energy Systems: Minimizing Emissions and Water Use and Assessing Response to Drought in the Texas Electrical Grid

David Allen, Chemical Engineering, University of Texas at Austin, Austin, TX

Electricity generation accounts for approximately 40% of energy use in the United States, and consequently, developing mechanisms for electricity generation that minimize impacts, yet are resilient to stress, is a key element in the development of sustainable energy systems. As a case study of the sustainability of electrical grids, this presentation will quantitatively examine trade-offs between air emissions, carbon intensity, and water use in the electrical grid operated by the Electricity Reliability Council of Texas (ERCOT). In addition, the response of the grid to drought stress, which is assumed to limit the availability of cooling water, and the response of the grid to changes in relative prices of natural gas and coal, will be reported.

(426f) Exergy Analysis of Solid Waste Management Opportunities in Indian Context

Aniruddha Pandit, Chemical Engineering Department, Institute of Chemical Technology, Matunga, Mumbai, India, Mumbai, India

Increase in population, urbanization, economic growth, Industrialization, and change in the lifestyle in India had resulted in to increase use of the resources. Increased use of the material resources and its inefficient conversion in to useful work results in generation of waste. A huge quantity of Municipal Solid Waste (MSW) gets generated each year. Most of the MSW generated is usually disposed of unscientifically in open dumps and landfills, creating problems to public health and the environment. As per the 2001 census around 1billion India's population lives in urban areas. MSW generation in cities in India is found toincrease from 6 to 48 million tons from year 1947 to 1997 with an annual growth rate of 4.25%. The present generation rate of MSW in Mumbai is estimated to be about 7500 tons while the state of Maharashtra as a whole generates about 16000 tons per day as estimated by a Planning commission report. In many cities more than half of the solid waste generated remains unattended. It was estimated that the MSW generation and disposal resulted in the emissions of 12.69 million tons of CO2 eq. in 2007. Management of Municipal Solid Waste generated in cities of India is a cause of worry as far as environmental degradation is concerned. On the other hand the MSW contains high exergy content materials, which makes MSW as a potential energy source. The traditional way of management of these MSW is by dumping these wastes in a low-lying area called a landfill. Various energy recovery treatment methods are used for treatment of MSW, such as incineration, gasification, plasma treatment etc. Incineration includes direct combustion of waste while gasification includes converting the waste into a combustible gas by a gasification agent. Exergy analysis has been found to be an appropriate measure for material resource accounting as well as for accounting of the waste generated. Exergy gets lost in all real processes in the form of heat or physically, chemically or Nuclear radiation reactive substances. Exergy lost in the form of chemically/physically reactive substances is found to have major role in the degradation of the environment. The aim of this work is to analyze the MSW management in India using exergy analysis. In this work we will compare the various MSW treatment methods (Incineration, gasification, landfill, plasma treatment, etc.) on the basis of the exergy of all resource consumption, the exergy of the product output (in the form of electricity, methane gas, etc.) and the emissions. This assessment will provide the exergetic improvement potential in each of the disposal methods used which will lead to minimization of losses, emission and hence the environmental degradation. Outcome of the this work is expected to provide the insights in the exergetic potential of the MSW in India, its potential to harm the environment as well as the potential to energy conversion which may help to frame the future policies for MSW management in the country.

(426g) Solid Electrolyte Technologies for Energy and Environment

Ashok Joshi, Ceramatec, Salt Lake City, UT

Ion conducting solids have potential applications in various Energy and Environmental Technology areas. While these applications are envisioned several decades ago, the technical maturity of selection and fabrication of devices using these materials suitable for commercial applications is finally beginning to emerge.

Ceramatec has over 35 years of technology development experience in Electrochemical devices and has entered into the commercial development phase by deploying engineered prototype for field testing and validation.

Ceramatec's development partners are global including Indian Corporations and Institutions. In this regard, the potential role and viability of these ionic technologies for the Indian market will be discussed.

A summary of current developments is described in this presentation.

(459a) Multiblock Polymers: Panacea or Pandora's Box?

Frank S. Bates, Department of Chemical Engineering and Materials Science, University of Minnesota, Minneapolis, MN

Synthetic polymers impact nearly every segment of the chemical engineering profession with applications in commodity and value added plastics and elastomers, medical and separation technologies, the microelectronics industry, and a host of products that facilitate energy conversion and storage. Recent advances in polymer chemistry provide economically tractable strategies for combining two or more types of macromolecules into multiblock polymers capable of generating almost any imaginable nanoscale structure endowed with remarkable properties. However, increasing the number (n) and type (k) of discrete blocks leads to an explosion in sequencing possibilities, each capable of producing myriad morphologies, similar to the dilemma faced when designing new proteins for biological application. For example, just three sequences are possible with $n = 3$ and $k = 3$ (ABC, ACB and BAC) while $n = 4$ and $k = 3$ results in 9 distinct linear combinations. A modest increase in molecular complexity, $n = 6$ and $k = 4$, results in 300 unique possibilities! What engineering strategies are available for selecting among the multitude of choices? This lecture outlines the basic principles underpinning multiblock polymer science and engineering including modern synthetic approaches, how molecular architecture influences structure and properties, contemporary characterizations methods, and the current state of predictive statistical theories. Several commercially successful examples will be used to highlight the unparalleled opportunities for creating new materials and products presented by this emerging class of polymers.

(464a) Energy & Sustainability – What It Means for Europe

Richard Darton, University of Oxford and European Federation of Chemical Engineers, Oxford, United Kingdom

The countries of Europe are varied in geographical location, political system, state of industrialization and energy needs. Even amongst the 27 countries of the European Union (population ~500 million) there are significant differences. Energy policy within the EU27 is therefore a good illustration of how a group of nations can plan to work together to address the key issues of energy and sustainability. The EU's *Energy 2020 strategy* promotes energy saving (20% by 2020), free market movement of energy, and security of supply. It also plans a significant technological shift in the direction of decarbonisation, with development and demonstration projects such as second generation biofuels, CO2 capture and storage, electricity storage and electro-mobility, and next-generation nuclear. Mechanisms to support this strategy are a mixture of carrot and stick: funded research and innovation projects such as FP7, the world's largest research project, and directives leading to regulation. The strategy will undoubtedly book successes, though whether these will be fast enough or go far enough to meet external pressures, is an open question.

(464b) The Energy-Water Nexus from an Industrial Perspective

Henry T. Kohlbrand, AIChE and Consultant, Rapid City, SD

Water is required to produce energy and energy is required to make use of water. This energy water nexus is playing an increasingly important role in site selection and process synthesis as well as impacting existing facilities. As people's need for water increases, for agricultural and personal use, more and more industrial water use will be under pressure. New energy options including carbon capture systems, fracking systems for natural gas production and some biofuel systems are increasing our water demand. New and creative process solutions to address the energy water nexus will need to be developed and chemical engineers will play a critical role.

(464c) IChemE Roadmap - Framing the Issues and Addressing the Challenges

Ed Daniels, Shell Projects and Technology, London, United Kingdom

What does society need; what are the desirable outcomes and how can chemical engineers work in partnership with others to make it happen? These were the key questions addressed in the *Roadmap for 21st Century Chemical Engineering* published by the Institution of Chemical Engineers (IChemE) on the 50th anniversary of its Royal Charter in 2007.

Five years on, IChemE has conducted a detailed review of the key challenges facing humanity. In an updated report, *Chemical Engineering Matters*, due for publication in November 2012, the Institution aassesses the role of the chemical engineer in delivering process and product solutions that maintain and improve quality of life. The challenges are well recognised: energy, food & nutrition, water, health and wellbeing, but the review also stresses that several key issues and concerns must to be taken into account. These include safety and sustainability as well as the need for international and multidisciplinary collaboration. Education, training and research needs are also assessed. The review clearly identifies the essential role of the fundamental aspects of chemical engineering science and the systems thinking upon which the discipline is founded. It also asks the profession to directly confront the opportunities presented by advances in biotechnology.

In his presentation, Ed Daniels will offer an overview of the report's key findings and recommendations. He will also assert that the capability of the professional chemical engineer to design, build, operate and manage process solution that will maintain or improve quality of life is shaped by external factors that lie beyond science and technology. Drawing on nearly 25 years experience in the global energy business Ed will make the case for stronger engagement by chemical engineers and their professional bodies in wider public policy debate.

(464d) The Role of Chemical Engineering Societies in Facilitation of Carbon Management

Jeffrey J. Siirola, Purdue University and Carnegie Mellon University, Kingsport, TN; Purdue University and Carnegie Mellon University, Kingsport, TN

The role of chemical engineering professional societies in maintaining active promotion and facilitation of carbon management advances are discussed. This is in the context of an environment of increasingly available and affordable fossil gas, global economic stress, and lack of political and economic drivers for greenhouse gas mitigation.

(464e) Decarbonization Options: Policies and Technologies

Kamel Bennaceur, Schlumberger, Rio de Janeiro, Brazil

Without major changes in energy-related policies, the world's energy demand will grow by more than 40 % over the next two decades, and will continue to be dominated by fossil fuels. The share of coal in the global energy mix will continue to grow, leading to an increase in CO2 and other greenhouse gas emissions. Projections for 2050 of those emissions would lead to temperature increases in the range of 4-7 degrees C., according to the 2007 Intergovernmental Panel on Climate Change Assessment Report.

Options for an "energy revolution" in both the energy demand and the energy supply sectors are presented. Scenarios for the deployment of a new energy portfolio are discussed, including ones that would lead to a reduction of emissions by 50 percent in 2050. Among technologies that would mitigate emissions from the supply side, CO2 Capture and Storage (CCS) is a major emission abatement option. The cost of implementing these scenarios is given, along with the implications on the use of fossil-fuels.

(492a) A Life Cycle Model of Water Use in India with Implications to Manufacturing

Bhavik R. Bakshi, William G. Lowrie Department of Chemical and Biomolecular Engineering, The Ohio State University, Columbus, OH and Shelly Bogra, TERI University, New Delhi, India

With India's fast economic growth, increasing prosperity, changing consumer preferences, and erratic monsoons, fresh water is becoming a critical resource that could greatly limit the country's development. While agriculture continues to be the largest direct user of water, other large direct consumers include the mining industry and thermoelectric power generation. The rest of the economy, including the chemical and manufacturing industries, depend indirectly on these primary water users, making it essential to understand the direct *and* indirect dependence. Such understanding is essential for determining the risk and vulnerability of economic activities to water scarcities and opportunities for enhancing water efficiency by managing the supply chain, and is becoming an important part of business decision making.

This presentation will describe a new model for determining the life cycle water withdrawal of all sectors of the Indian economy. This model combines data about direct water use in India with an economic input-output model of the Indian economy. The data about water use was collected from diverse sources including government statistics, industry sustainability reports, and the academic literature and combined with a model of the 2005 Indian economy. In addition to providing the life cycle water withdrawal for all sectors, this model can also be used to determine the supply and demand chains for each sector and to identify “water hotspots” or sectors in the selected network that are most dependent on water. Such information can help by guiding industrial decisions and government policies that reduce or manage the risk of scarcities, and for improving enhancing efficiency. In addition to providing novel insight about the life cycle withdrawal of water in India, this work will focus on the chemical and manufacturing sectors, and identify vulnerabilities due to water withdrawal. These results will be compared with those for the United States economy. On-going research on modeling and understanding the water-energy nexus in India will also be described.

(492b) Energy for Sustainability

Ram B. Gupta, Division of Chemical, Bioengineering, Environmental and Transport Systems, The National Science Foundation, Arlington, VA

Today's world is facing two critical challenges: (1) high fuel prices, and (2) climatic changes. Both of these are linked to the overdependence on the fossil fuels: petroleum, natural gas, and coal. The transport is almost totally dependent on fossil particularly petroleum based fuels such as gasoline, diesel fuel, liquefied petroleum gas, and natural gas. The combination of rising oil prices, issues of security, climate instability and pollution, and deepening poverty in rural and agricultural areas, is propelling governments to enact powerful incentives for the use of renewable energy.

In this talk, NSF's Energy for Sustainability program will be presented. This program supports fundamental research and education that will enable innovative processes for the sustainable production of electricity and transportation fuels. Processes for sustainable energy production must be environmentally benign, reduce greenhouse gas emission, and utilize renewable resources. Projects include those related to biofuels, photovoltaic solar energy, wind energy, and advanced batteries for transportation. co RttP~H sdes such as jatropha oil and algal oil to obtain pure hydrocarbons, mostly n-alkanes. We have developed a process for biojet fuel production from jatropha oil and other triglyceride feedstocks. The long-chain hydrocarbons produced from hydrotreating reaction are selectively hydrocracked and hydroisomerized using suitably designed catalysts to give products that include jet fuel range hydrocarbons. Quality control specifications and approval protocol have been created to ensure that aviation biojet fuels can be directly use as a drop in fuel in the aircraft. Preliminary characterization of biojet fuel is done as per ASTM 1655 specifications. More detailed tests are being done to determine fit for purpose properties, followed by further tests such as those for fuel and engine system components so as to meet ASTM D 4054 specifications prior to final approval for use as a drop-in fuel[2].

References:

1. G. Marsh, Biofuels: aviation alternative, Renewable Energy Focus, 2008, 9, 48-51.

2. L. Rye, S. Blakey, C. W. Wilson, Energy and Environment Science, 2010, 3, 17- 27

(492c) Fungible Transportation Fuels from Bio Resources: Production of Jet Fuel from Non-Edible Low Cost Vegetable Oils

Madhukar Garg, Indian Institute of Petroleum, Dehradun, India

Jet fuels must meet very stringent international specifications, due to which it is difficult to develop an alternative fuel for aviation as compared to automobile applications[1]. There is no work reported till-date on designer catalysts that would maximize the conversion of triglycerides (specifically those from non-edible jatropha and algae oil) into jet-fuel range (kerosene) hydrocarbons. A single step economical process for producing aviation fuel from lipids has been inconceivable till-date. Triglyceride derived kerosene with its low aromatics content and very low sulfur content is also commercially very attractive for its numerous applications as solvent or other value added applications.

Plant-oil triglycerides are made up of one glycerol unit and three long-chain fatty acids. The carbon chain length and number of double bonds in the fatty acid vary depending upon the source of the oil. These can be hydrotreated and converted into pure hydrocarbons. Hydrotreating catalysts supported on silica-alumina and alumina supports were used to convert triglycerides such as jatropha oil and algal oil to obtain pure hydrocarbons, mostly n-alkanes. We have developed a process for biojet fuel production from jatropha oil and other triglyceride feedstocks. The long-chain hydrocarbons produced from hydrotreating reaction are selectively hydrocracked and hydroisomerized using suitably designed catalysts to give products that include jet fuel range hydrocarbons. Quality control specifications and approval protocol have been created to ensure that aviation biojet fuels can be directly use as a drop in fuel in the aircraft. Preliminary characterization of biojet fuel is done as per ASTM 1655 specifications. More detailed tests are being done to determine fit for purpose properties, followed by further tests such as those for fuel and engine system components so as to meet ASTM D 4054 specifications prior to final approval for use as a drop-in fuel[2].

References:

1. G. Marsh, Biofuels: aviation alternative, Renewable Energy Focus, 2008, 9, 48-51.

2. L. Rye, S. Blakey, C. W. Wilson, Energy and Environment Science, 2010, 3, 17- 27

(492d) Renewable Chemicals: Opportunities and Challenges

Bala Subramaniam, Department of Chemical and Petroleum Engineering, Center for Environmentally Beneficial Catalysis, University of Kansas, Lawrence, KS

The *only sustainable option* for producing chemicals in the long term is *plant-based biomass*. New chemistries and sustainable catalytic technologies are needed to produce chemical intermediates from plant-based feedstocks such as sugars, sugar alcohols and vegetable oils. This grand challenge provides fresh opportunities for chemists and chemical engineers to implement biorefineries that co-produce fuels and chemicals. The challenge is not different from what the petroleum industry faced nearly a century ago and has the potential to spur a new manufacturing sector in agro-based economies such as the United States and India. In recent years, there has been increased interest to source many megaton chemical intermediates (such as ethylene oxide, linear aldehydes and dicarboxylic acids including terephthalic acid) from biomass-derived feedstocks. Existing technologies for these commodity chemicals are waste generating and energy-intensive. Hence, to preserve the "green" potential of the renewable feedstocks, new sustainable conversion technologies are needed.

This talk will present examples of new, sustainable technology concepts for some of the aforementioned chemical intermediates. Integrated R&D approaches that fully exploit the synergies between catalyst design, tunable reaction media (such as supercritical fluids, gas-expanded liquids, ionic liquids) and novel reactor concepts (integrating reaction and separation) to develop such alternate technologies will be presented. The significant role of *quantitative* sustainability analyses in guiding the development of these processes toward commercialization will also be discussed.

(492e) Modelling Granular Flow Processes

Devang V. Khakhar, Department of Chemical Engineering, Indian Institute of Technology, Bombay, Mumbai - 400076, India

Handling granular materials is carried out on a large scale and usually involves significant energy inputs. The flow behavior is complex and an understanding of the flow would be useful for a better description of several industrial processes leading to more energy efficient designs. Flowing mixtures tend to segregate, altering the flow characteristics and introducing an additional level of complexity.

Several approaches have been successful in the description of flowing granular mixtures. We present continuum models for granular rheology and segregation in dense flows. The models are developed based on an understanding of particle level processes obtained from DEM numerical simulations (analogous to molecular dynamics) and experiments, both in model flows. Results from these studies will be reviewed. The models would be useful for design and scale of industrial processes involving dense flows.

(492f) MAGIC (Modular, Agile, Intensified and Continuous) Processes and Plants for Specialty Chemicals

Vivek V. Ranade, Cepd, National Chemical Laboratory, Pune, India

In recent years, process intensification based on micro-devices has been in vogue. Despite significant research and several efforts in commercializing these, the adoption of micro-devices based technologies by specialty chemicals industry is (at the most) slow. The hype surrounding micro-reactors was followed a period of disillusionment. This note highlights the need to look beyond micro-reactors and micro-devices for realizing process intensification in specialty chemicals industry. The note outlines a new effort based on MAGIC (modular, agile, intensified and continuous) processes and plants. It also highlights the need for involving industry as well as large number of students (chemical engineering and process chemistry) in this ambitious program of MAGICally transforming the way specialty chemicals are manufactured.

(492g) The National Programme on Technology Enhanced Learning: Beyond Open Educational Resource Creation and Without Borders

Mangala Sunder Krishnan, Department of Chemistry, I.I.T Madras, Madras, India

The National Programme on Technology Enhanced Learning is an academic initiative by the Indian Institutes of Technology and the Indian Institute of Science with the objectives of improving the quality of engineering and science education in India and to enhance the employability of a large number of graduates emerging from engineering and science institutions in India. The outcomes from the project are full length (semester-long) courses on all major engineering and science subjects at the undergraduate and postgraduate levels at present with a considerable number of video (>10,000 one-hour records so far) and web based supplements and references for further reading, all made available in a free and open form under the Creative Commons License. Institutions all over the world are encouraged to receive the repository free of charge and use them without commercial intent. In addition, NPTEL is establishing collaborations with Google and a number of industries in India, and educational networking private partners such as Classle Inc. and is open to faculty, subject-matter experts outside of the IIT / IISc system for creating the best educational learning resources, which are also peer reviewed and authenticated. NPTEL in its further phases, will extend invitation to faculty abroad to contribute either as course modules or as special topics that would be useful for postgraduates and young researchers all over the world. It is also being extended to management, arts and humanities and several other disciplines. A very large effort is considerably underway to study the impact of the programme and to get user feedback from both educational and industrial users and to offer valuable certification programmes to help the young graduates in their career prospects. NPTEL is solely funded by the Ministry of Human Resource Development, Government of India and has received about 22 million US dollars (Rs. 116 crores) so far.

The complete repository is available from http://nptel.iitm.ac.in.

The video repository is mirrored also in http://www.youtube.com/iit.

(501a) Programming Functions in Hierarchical Hybrid Nanostructures: Role of Chemical Engineers

Kookheon Char, School of Chemical & Biological Engineering, Seoul National University, Seoul , South Korea

Polymer thin films and nanohybrids with functions have recently received much attention due to many important potential applications such as nanolithography and nanopatterning for nanodevices, biologically relevant surfaces for sensors and drug delivery, photonic or electronic devices, and many others. Many nano-objects (such as quantum dots, nanotubes or nanowires, and nanosheets) carrying unique functions should be incorporated into the polymer matrices to achieve such practical goals. Tremendous efforts have also been devoted to the intelligent hybrids for energy efficient optoelectronic devices with multifaceted approaches such as noble design and synthesis of nanomaterials, optimal processing to realize ideal nanostructures, and their relationship with device performance.

In this presentation, I will discuss recent progresses in block copolymer thin films and layer-by-layer thin film depositions to program (multiple) functions within the polymeric films. Also, the hybridization schemes for organic-inorganic structures optimal for various applications will be presented and the hierarchical structure mimicking the Nature, combining both the bottom-up and the top-down approaches, will be briefly discussed

Considering the wide application areas of both functional thin films and hybrid materials, the approaches taken in the present study are likely to open up new possibilities for chemical engineers to realize new advanced materials or devices with multifunctional properties.

(501b) A Forty-Five Year Transformation From Engineer to Scientist – What Took So Long?

Chang Dae (Paul) Han, Polymer Eng, University of Akron, Akron, OH

After forty-five years in academia, it is only within the past few that I have been able to think of myself as a scientist. I began my academic career at Polytechnic Institute of Brooklyn in 1967, after completing my graduate studies in Chemical Engineering at MIT and then three years in industry. The first fifteen years of my research activities were mostly experimental in nature, focused primarily on developing a fundamental understanding of relationships between the viscoelastic properties of molten polymers (polymer rheology) and their processability in various polymer processing operations by the application of chemical engineering principles. To really understand the influence of the molecular characteristics of molten viscoelastic polymers on their physical properties required, I had to study statistical physics in order to understand the molecular theory of entangled polymer melts. After several years of applying theory to commercial polymers for my research activities, the inability to control their molecular parameters became a limiting factor to progress. Nearly midway into my academic career, I began the next phase as polymer chemist in which I began to synthesize my own block copolymers and subsequently liquid-crystalline polymers. It turned out that these polymer structures had complicated morphology and phase transitions that could be induced by flow and temperature. In 2003 I began to synthesize *functional* block copolymers and *functional* liquid-crystalline polymers having chemical structures which had not yet been reported in the literature. Using these functional polymers I investigated the effectiveness of the functional polymers on the preparation of exfoliated nanocomposites. A new, totally unanticipated, turning point of my academic research activity began when in 2007 I began efforts to try to synthesize molecule-based (impurity-free) magnetic polymers. There are two factors which have motivated me to undertake this challenge. The first is that no one has ever reported on successful synthesis of magnetic polymers although numerous research groups had tried for over 50 years. The second is that the synthesis of magnetic polymers demands a good understanding of electronic structure theory based on quantum chemistry. During past five years I had to synthesize not only new organic compounds and subsequently magnetic polymers, but also study quantum chemistry for a better understanding of the chemical reactivity of two functional compounds. In short I transformed myself into an organic/polymer chemist over the course of forty-five years. Although I retired this past May, as an Emeritus Professor I continue to synthesize magnetic polymers and magnetic conducting polymers, and also conduct numerical computations, incorporating electronic structure theory, of the molecular orbital of the functional monomers and magnetic polymers which I have synthesized in my laboratory, which I hope to continue for many years to come.

(501c) Progress in Tissue Engineering

Kristi S. Anseth, Department of Chemical and Biological Engineering, University of Colorado, Howard Hughes Medical Institute, Boulder, CO

Tissue engineering is an interdisciplinary field that marries the principles of engineering and the life sciences toward the development of biological substitutes that can restore or improve tissue function. Typically, tissue engineers use advanced biomaterials that serve as scaffolds for the delivery of cells, and the scaffolds further provide cues and signals to the cells to guide their re-growth or healing of a tissue. One of the products of tissue engineering, a temporary skin substitute, is being used to treat patients with burns and diabetic ulcers. This talk will provide three examples to illustrate the complexities of designing tissue engineering strategies to: 1. heal damaged cartilage, 2. engineer living heart valves, and 3. treat Parkinson's disease and other neurological disorders. Finally, all of this will be placed in the context of the many scientific and regulatory challenges that exist and provide insight related to the future promise of this field in clinical medicine.

(501d) Thin Film Solar Cells From Nanocrystal Inks of Quaternary Semiconductors

Rakesh Agrawal, School of Chemical Engineering, Purdue University, West Lafayette, IN

The production of low-cost solar cells from earth abundant elements is necessary for the use of solar electricity on a large scale. The creation of a suitable inorganic colloidal nanocrystal ink for use in a scalable coating process is a key step in the development of low-cost thin film solar cells. We have developed an innovative method of using copper indium gallium disulfide (CIGS) and Cu_2ZnSnS_4 (CZTS) nanocrystals as the building block for the fabrication of bulk absorber thin films. The nanocrystal ink solution is applied directly on various substrates to form a thin film coating. The nanocrystals are then consolidated into large crystalline chalcopyrite domains by a brief thermal treatment under Se vapor. Furthermore, the ability to control the composition of nanocrystals allows the unique capability to bandgap engineer the selenized absorber layer.

In this talk we will present some of our recent results on CZTS nanoparticle synthesis, absorber film sintering and optoelectronic characterization of the resulting solar cells. These results are exciting as they provide us the insights that will eventually lead to highly efficient low-cost solar cells.

(567a) Sustainable Energy Research in Chemical Engineering

Rakesh Agrawal, School of Chemical Engineering, Purdue University, West Lafayette, IN

Historically chemical engineering has played a central role in the current fossil resource driven world. Now, as the human race prepares to make inevitable transition from the fossil resources, chemical engineering is well poised to contribute to this transition.

Here we will discuss some of our sustainable energy related research. An energy systems analysis, to identify opportunities and pitfalls as the renewable economy emerges, will be presented. We will focus on a future where the basic human needs of food, chemicals, heat, electricity and transportation will generally be met by solar energy. In a solarenergy- driven world, it will be particularly challenging to satisfy the need of the transportation sector due to its requirement of high energy density fuel and associated ease of handling. Some novel solutions to meet this challenge and sustain the current transportation sector will be presented. The importance of developing transition solutions that use both hydrocarbon resources as well as solar energy in a synergistic manner will be emphasized through examples.

Another important challenge relates to energy storage at Tera Watt-hours scale. The intermittency of renewable resources such as solar energy and wind necessitate solutions that will supply on demand energy around the clock. We will present some of our initial solutions to answer this challenge. Such opportunities indeed make the future of chemical engineering exciting.

(567b) An Overview of the Decatur Illinois CO2 Capture Projects

James J. Foster, Archer Daniels Midland Company, Decatur, IL

The Illinois Basin is hosting several major carbon capture projects. This basin, which underlies most of the state of Illinois, parts of Kentucky, and Indiana, ranks among one of North America's the best sites for potential storage of anthropogenic CO_2 emissions. Within this basin, the Mount Simon Sandstone, a major regional saline reservoir, is the target for sequestration because it has good permeability and porosity with overlying strata of impermeable shale. Because the regional thickness of this reservoir increases towards the center of the basin, the optimum location for maximum storage of CO_2 is in north central Illinois. Because of the excellent regional geology and access to industrial scale quantities of CO_2, two projects are being conducted at the Archer Daniels Midland Company's (ADM) agricultural processing and biofuel production facility located in Decatur, Illinois. Both projects will demonstrate the ability to inject and store industrial scale quantities CO_2 emissions into the Mount Simon Sandstone.

Illinois Basin - Decatur Project (IBDP), (**Status- in Operation**). This project is led by Illinois State Geological Survey (ISGS), under the Midwest Geological Sequestration Consortium (MGSC) Regional Carbon Sequestration Program, and is a large-volume, saline reservoir sequestration test that will inject approximately 333,000 metric tons of carbon dioxide (CO_2) per year for three years.

Illinois Industrial Carbon Capture and Storage Project (IL-ICCS), (**Status- under Construction**). This project is led by ADM and will expand the sites CO_2 injection and storage capability to that of a commercial-scale operation. The project is scheduled for startup in the third quarter of 2013 and will inject up to 1.0 million metric tons per year over an operational period of approximately 2.5 years. ADM will integrate the IBDP compression and dehydration facilities with the new facilities constructed under the IL-ICCS project upon completion of IBDP injection operations in fall 2014. A significant benefit of these two complimentary projects is the unique opportunity to better understand the interaction between the CO_2 plumes and pressure fronts emanating from two injection wells in the same sandstone formation.

IL-ICCS project has the following objectives:

- Demonstrate an integrated system for collecting CO_2 from an ethanol production plant and geologically sequestering in a saline sandstone reservoir:
- Conduct required geologic site surveys, site characterization and modeling.
- Design, construct, and operate a new CO_2 collection, compression, and dehydration facility capable of delivering up to 2,000 metric tons of CO_2 per day to the injection site.
- Integration of the new facility with an existing 1,000 metric tons per day CO_2 compression and dehydration facility to achieve a total CO_2 injection capacity of 3,000 metric tons per day or one million tons annually.
- Design, construct, and operate a storage site capable of accepting up to 3,000 metric tons of CO_2 per day.
- Implementation of deep subsurface and near-surface monitoring, verification, and accounting (MVA) plans for the stored CO_2.
- Develop and conduct an integrated communication, outreach, training, and education initiative.

The IL-ICCS project has leverage the knowledge and experience gained during the IBDP project. Site selection, reservoir modeling, MVA development, risk assessment, community outreach, engineering design, and facility construction are many of the areas in which the project team benefitted from the experience and lessoned learned during the IBDP project. Because of this experience, the IL-ICCS project has an accelerated design and construction schedule and plans for operation within 24-30 months.

Thermodynamic Modeling of Coal gasification : A Universal Approach

S.Pushpavanam and T. Renganathan

Chemical Engineering Department

I.I.T Madras

Chennai 600036

Email :spush@iitm.ac.in;renga@iitm.ac.in

Introduction

Modeling of a gasifier can be done using two approaches viz, thermodynamic or equilibrium modeling and rate based modeling. Equilibrium modeling is based on chemical and phase equilibria. This is valid when the reaction rates are sufficiently high so that we can assume that equilibrium is reached instantaneously. This assumption is valid for sufficiently high temperatures and high residence times. Under these conditions the exiting streams can be assumed to be in equilibrium. This approach however cannot be used to design a gasifier i.e determine the length and diameter of the gasifier or flow-rate of the different streams. It assumes that rates of reactions are faster relative to the residence time in the reactors. This approach has its advantages in the fact that it does not require the reaction rates to be available , however it is not suitable for detailed design of gasifiers.

On the other hand kinetic modeling takes into account hydrodynamics, transport processes and reaction kinetics. The primary drawback of this approach stems from the fact that the reactor is a multiphase system and most of these interactions are not well understood. . While the poor understanding of these processes is a limitation, this approach can provide insight in sizing and designing a gasifier. The reliability of this approach increases when the kinetics as well as the transport processes are well understood.

In this work we discuss the equilibrium approach to modeling in detail. One of the primary advantages of this approach is it does not require information of the reaction rates of the reactions which occur in the system. In the coal gasification process several reactions go on simultaneously i.e., the combustion reaction, the gasification reaction etc.The heterogeneous combustion and gasificaton reactions which occur in the system are dependent on the feedstock: is it from biomass or coal. This broad classification helps identify the overall composition of the feedstock. Within each type of the feedstock there can be significant differences in the reaction rates. The carbonaceous fuel is hence a non conventional feed. . Hence to use the kinetic modeling approach to design gasifiers would need reaction rates of the feed stock. This critical information may not be available for most feed stocks. In this system an additional challenge is the fact that the kinetics of the reaction is specific to the coal being used. For instance it depends on the morphology of the coal, its composition. Thus coal from one coal field will have a different set of kinetic constants compared to that from another coal field. The equilibrium approach does not need kinetic information and this is hence a big advantage.

The thermodynamic equilibrium model can be based on two approaches. It will be shown that these two approaches are equivalent and give the same results. In the **stoichiometric approach** the information of possible reactions occurring in the system is needed. This approach is based on a set of selected independent reactions. It uses elemental balances which ensures that mass is conserved and considers the reactions to be at equilibrium. One challenge is that the number of reactions which need to be specified and the nature of the reactions which need to be specified can vary as we change the operating conditions. We will see this in detail later.

The **non - stoichiometric approach** (also called Gibbs free energy minimization approach) is based on a set of selected species assumed to be present in the syngas which exits the gasifier. This method minimizes the total Gibbs free energy of the system and is subject to the constraints of elemental balances which ensures mass is conserved. It also ensures that energy is conserved.. The fact that this approach needs only knowledge of species expected to be present in the product stream irrespective of the reactions occurring in the process makes it advantageous over the stoichiometric approach. An added advantage of this approach is that we see the presence of various species in the exit gas. Should the species be not present in the system the approach will yield the concentration of the species to be zero. Hence no special modifications have to be done while using this approach to take into account that some species may not be present under some operating conditions. . Advantages of this approach will become clear when it is applied to the coal gasification process.

Equilibrium modelling

In the present work, the fuel is assumed to be dry (moisture free). Typical feedstocks contain carbon, hydrogen, oxygen, nitrogen, and sulfur. In addition to this the feedstock will contain ash, essentially inorganic matter. To begin with the fuel is assumed to contain only carbon, hydrogen and oxygen. This assumption is relaxed later. Under these assumption the fuel is represented as $C_{n_C} H_{n_H} O_{n_O}$. The justification for taking these three elements are that these three elements make up to about 85 to 90 % of most fuels. The results and conclusions obtained when only these three elements are considered do not deviate much when other elements are also included. In other words the presence of the other elements do not significantly affect the results. Secondly considering only these three elements enables us to uniquely represent the fuel and present the results as contour plots in the *H/C* and *O/C* ratio plane. This depiction of results allows us to present the results for any fuel. Hence the results have a universal characteristic.

In a typical industrial process the gasifying agent used is either oxygen or air or steam or carbon-dioxide.. The components assumed to be present at equilibrium are $C_{(s)}$, CO, H_2, CO_2, H_2O and CH_4. These are the components in the exiting synthesis gas.

The global gasification reaction can be written as

$$C_{n_C}H_{n_H}O_{n_O}N_{n_N}S_{n_S}.wH_2O + mO_2 + 3.76mN_2 + sH_2O \rightarrow n_{C(S)}C(S) + n_{H_2}H_2 + n_{CO}CO + n_{CO_2}CO_2 + n_{CH_4}CH_4 + n_{SO_2}SO_2 + n_{N_2}N_2 + n_{H_2O}H_2O \quad \text{(R2.1)}$$

Here $s = 0$ ($m = 0$) for gasification using only air (only steam) respectively. Our objective is to determine the coefficients n's which occur on the right hand side of the reaction R 2.1.The right hand side of this equation should contain all the species that are expected because of the gasification. Thus for example if ammonia or hydrogen sulfide are expected in the gaseous stream they should be included as possible products.

The conservation of mass as applied to the system gives rise to the elemental balance equations. For each element balancing the number of atoms in the overall reaction, we obtain:

Element C balance: $$n_C = n_{CO} + n_{CO_2} + n_{CH_4} + n_{C_{(S)}} \quad \text{(R2.2)}$$

Element H balance: $$n_H + 2s = 2n_{H_2} + 4n_{CH_4} + 2n_{H_2O} \quad \text{(R2.3)}$$

Element O balance: $$n_O + s + 2m = n_{CO} + 2n_{CO_2} + n_{H_2O} \quad \text{(R2.4)}$$

The above balances state that the number of atoms of each element entering is the same as the number of atoms of each element leaving the system.

The energy balance is,

$$H_{prod}(T) = H_{react}(T_I) + Q \quad (2.1)$$

where,

$$H_{react} = \sum_{react} n_i H^0_{f_i,298} + \sum_{react} n_i \int_{298}^{T_i} C_{pi} dT \qquad (2.2)$$

$$H_{prod}(T) = \sum_{prod} n_i H^0_{f_i,298} + \sum_{prod} n_i \int_{298}^{T} C_{pi} dT \qquad (2.3)$$

Enthalpy of fuel is calculated using,

$$H_{fuel} = M_{fuel} HHV - n_C H^0_{f_{CO_2},298} - \frac{1}{2} n_H H^0_{f_{H_2O},298}$$

(2.4)

Higher heating value (*HHV*) of the fuel is calculated using the correlation proposed by Channiwala and Parikh (2002)

$$HHV = 0.3491\, x_C + 1.1783\, x_H - 0.1034\, x_O$$

(2.5)

The range of validity of the above correlation is $0 < H/C < 1.81$ and $0 < O/C < 0.83$. The correlation has been developed based on specific *H/C* and *O/C* combinations (for the different feedstocks in Van Krevelen diagram). In this work, this correlation is assumed to be valid for all combinations of *H/C* and *O/C*.

Non stoichiometric approach

For a given pressure, the moles of the different components and temperature are determined such that the total Gibbs free energy is minimum subject to the elemental balances, as given below,

The non-stoichiometric model is based on the concept that at equilibrium condition, the total Gibbs free energy of the system is minimum (Smith et al., 2003). The total Gibbs free energy is a function of temperature, pressure and the number of moles of ith component (n_i) present at equilibrium. Hence, the total Gibbs free energy of the system is given by

$$G^t = f(T, P, n_{H_2}, n_{CO}, n_{CO_2}, n_{CH_4}, n_{H_2O}, n_{C_{(S)}}) \qquad (2.11)$$

At equilibrium, since the total Gibbs free energy is minimum.

$$dG^t = 0 \qquad (2.12)$$

The problem is to find the set $\{n_i\}$ and T which minimize the G^t for given T and P, subjected the constraints of the elemental balance.

The simulation is carried out using Aspen plus based on non-stoichiometric approach. The Aspen Plus flowsheet used for simulation is shown in Fig 3.1.

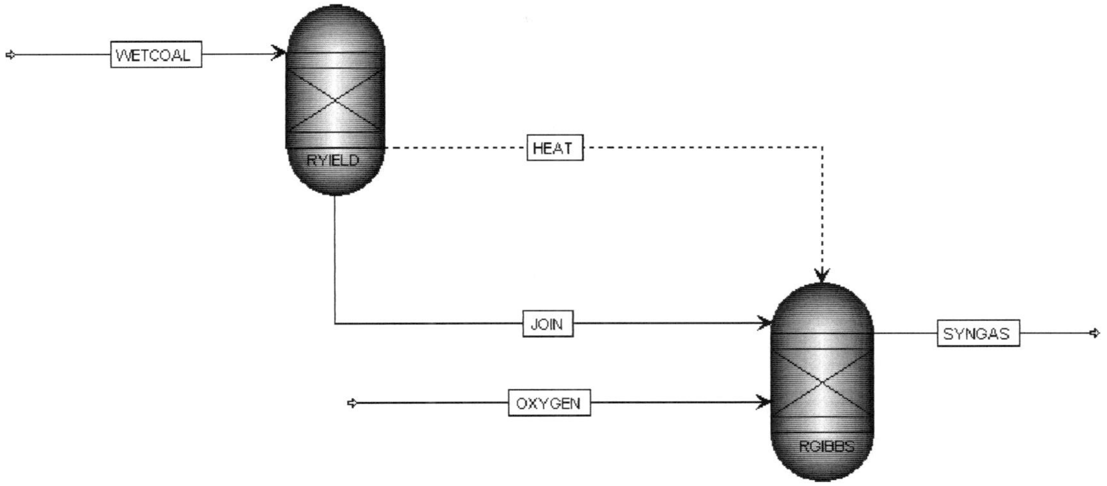

Fig. 1 Aspen Plus flowsheet for gasifier

The RYIELD and RGIBBS reactors are used to simulate the gasification process. The RYIELD reactor is used to 'split' the fuel into its component elements. This is necessary since the feed-stock is a non conventional. The RYIELD output consists of elements which can be fed to the next unit the RGIBBS. The RGIBBS reactor is used to perform multiphase - reaction equilibrium calculation based on the Gibbs free energy minimization approach. (HOW DOES ASPEN DO TIS) . Adiabatic operation of the system is ensured by giving the heat output of RYIELD reactor as heat input to RGIBBS reactor (This is shown as a dashed line). The sensitivity analysis feature of Aspen Plus is used to study the effect of various combinations of *H/C* and *O/C* ratios. For a specific *H/C* and *O/C* atomic ratios, the ultimate analysis in terms of mass percentage of the elements carbon, hydrogen and oxygen is determined. In each case the total fuel flowrate is kept constant. For each case, a design specification of carbon entering equal to carbon leaving in gaseous species is used to determine the oxygen flow rate required for complete conversion of carbon (carbon boundary point).

The fuel is specified as a non conventional component. The enthalpy model used for coal is HCOALGEN with user supplied heat of combustion and the density model is DCOALIGT.

Two calculator blocks are used.(SHOW SCREN SHOTS OF THIS) The first calculator block evaluates the heat of combustion of feedstock using the correlation proposed by Channiwala and Parikh (2002) for HHV. The second calculator block is used to specify yield fractions for RYield reactor using the ultimate analysis data. Ultimate analysis of lignite coal used in simulation is given in Table 3.1. The specifications for coal and gasifying agent streams are given in Table 3.2.

Table 1 Ultimate analysis of coal

Components	Weight %
Ash	0
Carbon	66.67
Hydrogen	6.67
Oxygen	26.66

Table 2 Gasifier inlet stream specifications

Stream and specification		Value
Fuel entering temperature (K)		298
Oxygen	Entering temperature (K)	298
	Entering Pressure (atm)	1

Results and Discussion: Gasifier performance is analyzed for *O/C* varying from 0 to 0.9 and *H/C* varying from 0 to 1.8, the limits obtained from the Van Krevelen diagram (Fig. 2.). Results are presented on gasification temperature, syngas composition (wet basis for water and dry basis for others). The results are presented as contours on Van Krevelen coordinates namely *H/C* vs *O/C* and the effect of *H/C* and *O/C* on the simulation results is analyzed in detail.

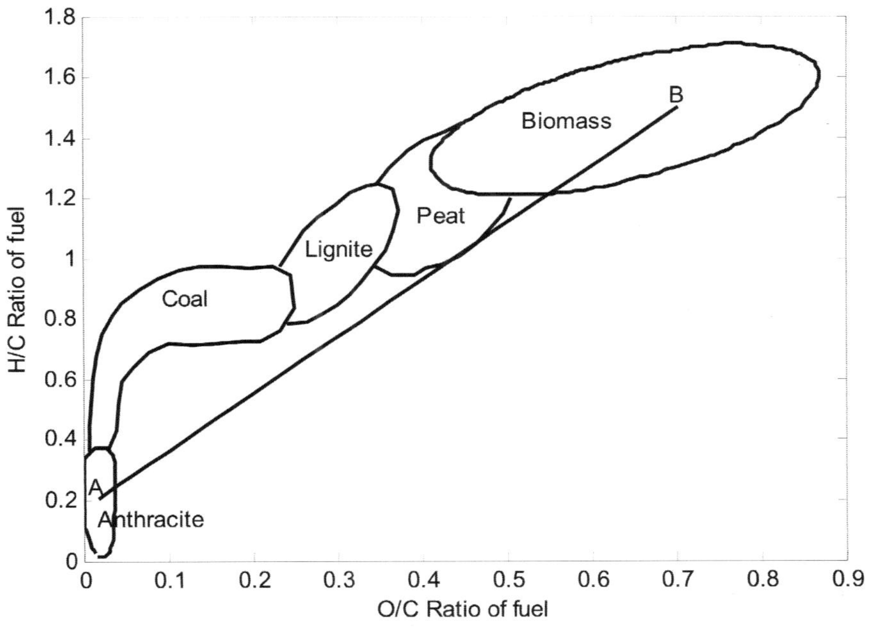

Fig. 2 Representation of different feedstocks on Van Krevelen (*H/C* vs *O/C*) coordinates

The different feed stocks as represented by regions in Van Krevelen diagram are also superimposed on the contour plots to assist in directly estimating the performance for different feedstocks. Co-gasification, which refers to using a mixture of two or more feedstocks (e.g. coal and biomass) as feed to the gasifier, results in *O/C*, *H/C* ratios of mixtures lying outside regions of single feed-stocks, as shown by line AB in Fig.2. The performance of co-gasification of binary feed stock mixtures is also discussed

Gasification using oxygen

Figure 3 depicts the dependence of gasification temperature, fraction of unconverted carbon, composition of synthesis gas and cold gas efficiency (CGE) as a function of oxygen flow rate (mole of oxygen per mole of carbon) for gasification lignite coal. It can be seen that, with increase in oxygen flow rate, the fraction of carbon unconverted decreases and for a particular oxygen flow rate (0.38 mole of oxygen per mole carbon) all carbon in the fuel gets converted and no carbon appears in the exit syngas. At this point mole fraction of CO is maximum and that of CO_2 is minimum and the *CGE* is maximum. This point is called Carbon Boundary Point (CBP) (Prins et al., 2003). Hence for maximum CGE the gasifier has to be operated at CBP. The generalized performance valid for any fuel is analyzed at CBP.

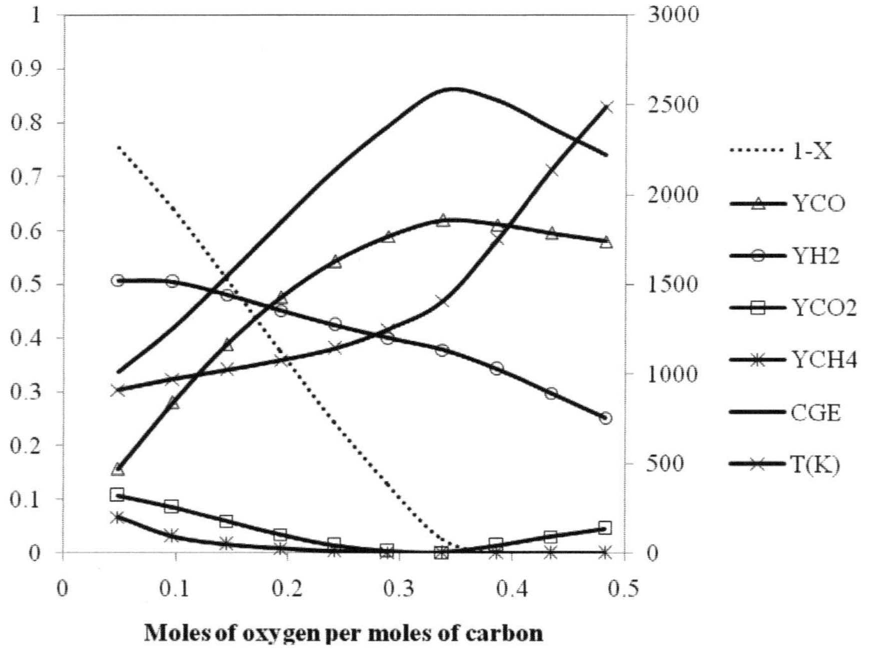

Fig. 3. Effect of oxygen flowrate on gasifier performance.

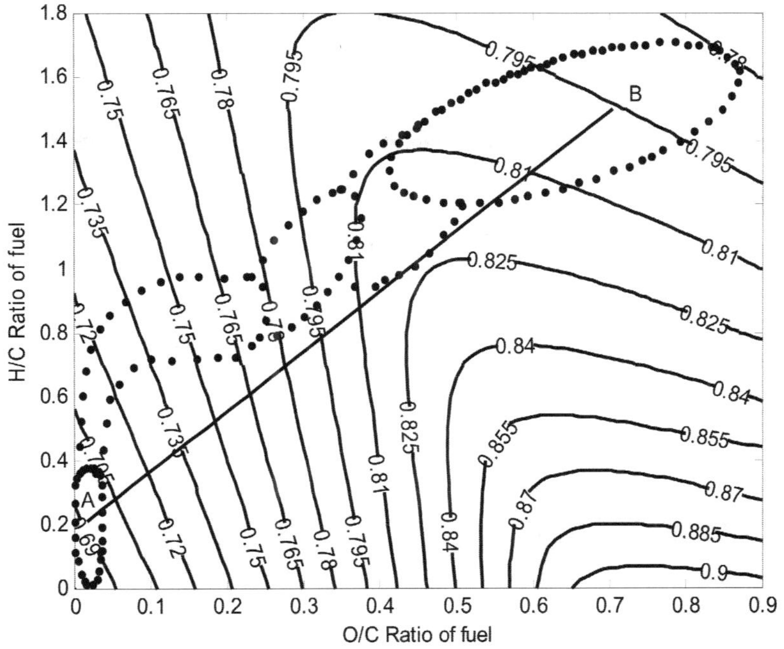

Fig.4 Contour plot of cold gas efficiency (CGE) for gasification of fuel using oxygen

The contour plot of cold gas efficiency (*CGE*) for gasification of fuel using oxygen at carbon boundary point is shown in Fig. 4.. For a given *H/C* ratio, the plot of *CGE* shows a maximum with respect to the *O/C* ratio. The shape of these curves indicates that there is an optimum combination of fuel mixtures which results in a maximum efficiency.

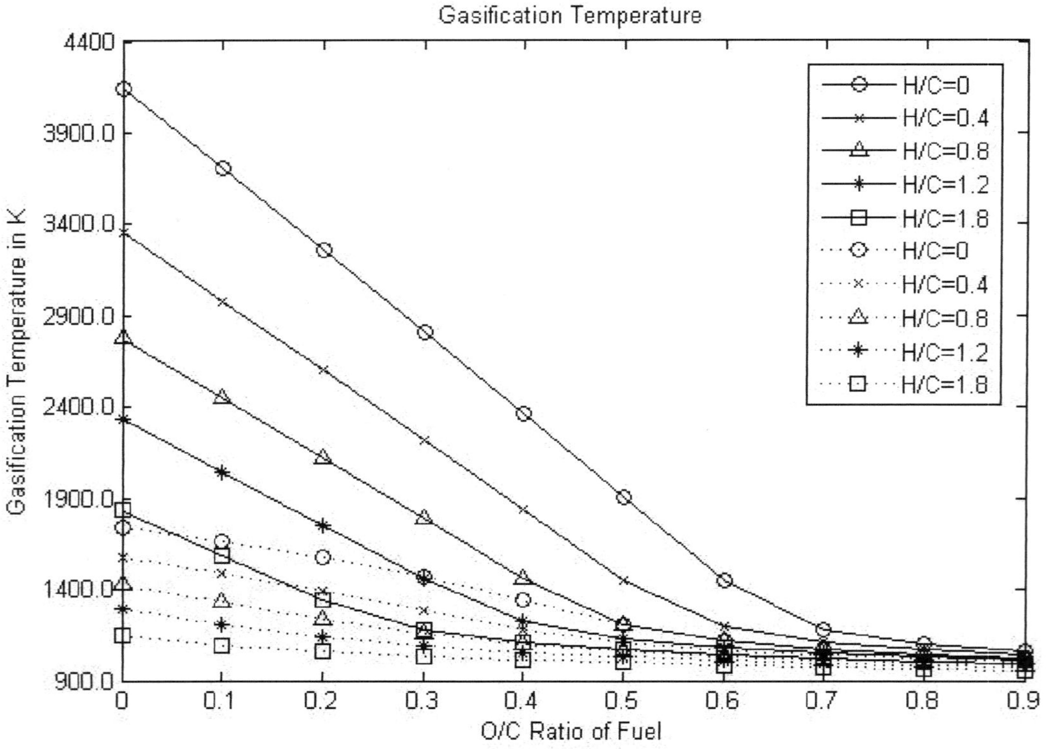

Fig. 5 Dependence of gasification temperature on O/C ratio of Fuel for various H/C ratios using air (solid lines) and oxygen (dashed lines) as a gasifying agent.

The temperature of gasification at carbon boundary point is greater in case of oxygen than Air, this is to be expected since nitrogen serves as an inert and its thermal capacity lowers the temperature. The cold gas efficiency is lower when air is used as the gasifying agent and the contour plots are shown in Fig.6.

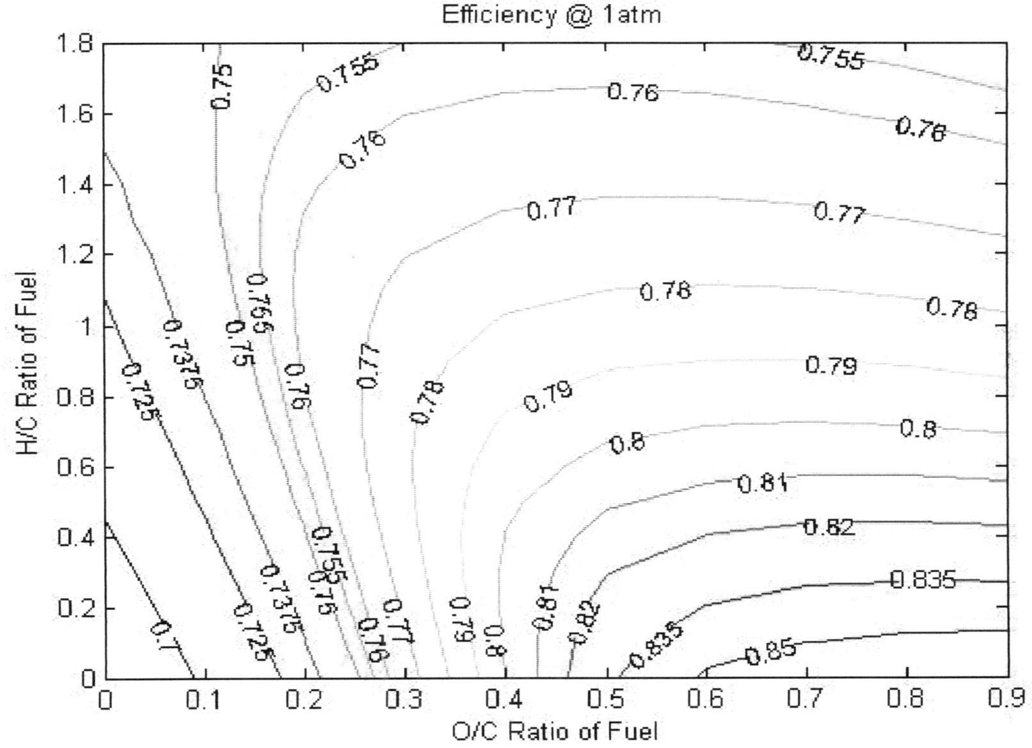

Fig. 6 Contour plot of cold gas efficiency (CGE) for gasification of fuel using air as gasifying agent at 1 atm pressure.

References

Benson, S. W., Thermochemcial Kinetics, Second Ed,; Wiley-Interscience, 1976.

Channiwala, S. A.; Parikh, P. P. A unified correlation for estimating HHV of solid, liquid and gaseous fuels. *Fuel* 2002, *81*, 1051-1063.

Ganesan V., Computer Simulation of Spark-Ignition Engine Processes; University Press, 1996.

Gorokhovski, M., Karpenko; Lockwood, E. I.; Messerle, F. C.; Trusov, V. E.; Ustimenko, A. B.; Plasma technologies for solid fuels: experiment and theory, *Journal of the Energy Institute*, 2005, 78(4), 157-171.

Kalinenko, R. A.; Kuznetsov, A. P.; Levitsky, A. A.; Messerle, V. E.; Mirokhin, Yu. A.; Polak, L. S.; Sakipov, Z. B.; Ustimenko, A. B., Pulverized coal plasma gasification, *Plasma Chemistry and Plasma Processing*, 1993, 13(1), 141-167.

Loison R. and Chauvin R., Pyrolise rapide du charbon, *Chimie et Industrie,* 1964, 91.

Prins, M. J.; Ptasinski, K. J.; Janssen, F. J. J. G. Thermodynamics of gas – char reactions: first and second law analysis. *Chem. Eng. Sci.* 2003, *58*, 1003-1011.

Ravikiran, A., Renganathan, T., Pushpavanam, S., Voolapalli, R.K., Cho, Y.S., Generalised Analysis of Gasifier Performance using Equilibrium Modelling, *Ind. Eng. Chem. Res.*, Article, 2011, ASAP.

Razak, A. M. Y. Industrial gas Turbine: performance and operability. Woodhead Publishing limited, Cambridge, England, 2003.

Reddy, B. V.; Ramkiran, G.; Kumar, K. A.; Nag, P. K. Second law analysis of a waste heat recovery steam generator, *Int. J. of Heat and Mass Transfer*. 2002, *35*, 1807-1813.

Smith, J. M.; Van Ness, H. C.; Abbott, M. M. Chemical Engineering Thermodynamics. Sixth Ed.; Tata McGraw – Hill Publishing Company Limited, 2003.

Stemmler, M.; Muller, M. Theoretical Evaluation of Feedstock Gasification Using H_2/C Ratio and ROC as Main Input Variables. *Ind. Eng. Chem. Res.* 2010, *49*, 9230-9237.

Vaezi, M.; Passandideh-Fard, M.; Moghiman, M.; Charmchi, M. Gasification of heavy fuel oils: A thermochemical equilibrium approach. *Fuel* 2011, *90*, 878-885.

Westbrook, C. K. and Dryer, F. L., Chemical Kinetic Modelling of Hydrocarbon Combustion, *Prog. Energy Combust. Sci.* 1984, 10, 1-57.

(567d) Chemical Looping Technology – Beyond CO2 Capture

L. S. Fan, Chemical and Biomolecular Engineering, The Ohio State University, Columbus, OH

The concept of chemical looping reactions has been widely applied in chemical industries, e.g., the production of hydrogen peroxide (H_2O_2) from hydrogen and oxygen using 9,10-anthraquinone as the looping intermediate. Fundamental research on chemical looping reactions has also been applied to energy systems, e.g., the splitting of water (H_2O) to produce oxygen and hydrogen using ZnO as the looping intermediate. Fossil fuel chemical looping systems had been applied to the steam-iron process for coal processing in the 1900s to the1940s and to the carbon dioxide acceptor process in the 1960s to 1970s. There are presently no chemical looping processes using fossil fuels in commercial operation. A key factor that hampered the continued use of these earlier processes for fossil energy operation was the inadequacy of the reactivity and recyclability of the oxygen carrying particles. This factor led to higher product costs for using the chemical looping processes, compared to using traditional techniques that processed petroleum or natural gas. With CO_2 emission control now being of a serious issue, interest in chemical looping technology has resurfaced. In particular, chemical looping processes are appealing due to their unique ability to generate a sequestration-ready CO_2 stream. Under the CO_2 constraint environment, the chemical looping technology clearly exhibits considerable advantage over traditional processes when accompanied by a CO_2 capture unit. However, if the CO_2 emission regulation is not enacted, under the CO_2 unconstrained environment, processes based on the chemical looping technology would require to be a standalone, commercially viable process without considering the benefit of the CO_2 capture.

This presentation will describe the fundamental and applied aspects of modern chemical looping technology that can convert fossil and biomass feedstock directly to various commercial products such as syngas, chemicals and liquid fuels. The chemical looping scheme linking to solid oxide fuel cell applications for high efficient electricity generation can also lead to an attractive chemical looping process option. The presentation will focus on reaction engineering issues associated with the complex metal oxide synthesis for performing selective oxidation reactions. Opportunities and challenges for chemical looping process scale-up and commercialization for these applications will also be illustrated.

(567e) Critical Sustainability Challenges in the Short Term

Jeffrey J. Siirola, Purdue University and Carnegie Mellon University, Kingsport, TN

Although there is much interest in concepts of sustainability within the chemical and energy industries, especially related to health, safety, and environmental performance, product stewardship, and efficient use of resources, the greatest sustainability challenge may well be that associated with fossil energy use which directly contributes to the majority of carbon dioxide emissions and its potential for significant climate change. It is now understood that reserves of fossil fuels are more abundant than previously believed, and it is very likely that they will continue to be exploited. In light of this situation, this presentation will discuss the challenges in mitigating greenhouse gas emissions from fossil fuel use, in substituting non-fossil energy, in substituting more energy intense fossil for another, in the design of more energy efficient chemical and power production systems, and new approaches to the optimum control of energy use during process operations.

(568a) Discovery to Delivery: The Role of Eulerian and Lagrangian Incentives in Translational Research

Sangtae Kim, Morgridge Institute for Research, Madison, WI

We are entering an era in which many universities are turning to translational research and the associated licensing revenue streams to make up for an anticipated shrinkage in traditional sources of funds. This shift in strategy is both encouraged and supported by recent actions of the major funding agencies. For the National Science Foundation's "Gen-3" Engineering Research Centers (ERCs), the award selection criteria includes vetting of the proposed plan for navigating the "valley of death" from research to commercialization. For the NIH, the new NCATS – the National Center for Advancing Translational Sciences, has the mission to "catalyze the generation of innovative methods and technologies that will enhance the development, testing and implementation of diagnostics and therapeutics across a wide range of human diseases and conditions."

These great aspirations for translational research face the reality of lackluster track records in technology transfer at most universities. Within this context, this presentation examines the incentive structures as a potential root cause for the less than stellar results (with the introduction of the key concepts of "Eulerian" and "Lagrangian" incentives) and offers several organizational strategies for universities that wish to venture into public-private partnerships and the formation of new institutes to boost translational research activities. The model will be illustrated with recent case studies involving medical devices and pharmaceuticals.

(568b) Theoretical Studies of the Effects of Structure and Composition On the Activity and Selectivity of Zeolite-Catalyzed Reactions

Alexis T. Bell, Department of Chemical and Biomolecular Engineering, University of California, Berkeley, Berkeley, CA

Zeolites widely used as catalysts for the conversion of petroleum to transportation fuels and for the synthesis commodity and specialty chemicals. The effects of the composition and structure of zeolites on their catalytic activity and selectivity have largely been revealed through experimental studies, but advances in theoretical methods and computer speed have opened the door for exploring the influence of zeolite composition and structure theoretically. This talk will demonstrate that a hybrid quantum mechanics/molecular mechanics (QM/MM) approach can be used to determine the energetic of species adsorption and formation of transition states with an accuracy approaching that of experimental measurements. QM/MM can also be used to determine the free energy of activation for elementary processes and the dynamics of reactions as they proceed towards product states. The latter type of calculations has revealed that multiple products can be formed from a given transition state and that the distribution of final products depends on a complex interplay of entropic and enthalpic effects. Examples of what can be learned from theoretical analysis of zeolite-catalyzed reactions will be illustrated with the cracking of alkanes, the methylation of alkenes by methanol, the oligomerization of alkenes, and the carbonylation of ethers.

SK Innovation: Present and Future

Byong-Sung Kwak, Global Technology, SK Innovation, 325, Exporo, Yuseong-gu, Daejeon, 305-712, Korea

SK Innovation is Korea's first and largest oil refining company and was established in 1962 as a national oil company. The ground breaking ceremony held fifty years ago in a small city of Ulsan signaled awakening and uprise of a 5,000 year old country of 'morning calm' from the aftermath of a dog-eat-dog war. For fifty years, our company has successfully fueled a world top ten economy and is doing business in the areas of oil exploration and production, refining, petrochemicals, polymers, lubricants, battery, and information and electronics materials with the annual sales revenue exceeding USD 60 B last year. The City of Ulsan now became Korea's seventh largest city and ranked first in terms of GDP per capita in Korea.

In this paper, we will go through how one company has grown by innovation and how it and its chemical engineers have been doing a pivotal role in changing a poor country's economy and people's quality of life dramatically.

(568d) Past, Present, and Future of Kiche

Seung Jong Lee, Chemical & Biological Eng, Seoul National University, Seoul, South Korea and Hyun-Ku Rhee, School of Chemical and Biological Engineering and Institute of Chemical Processes, Seoul National University, Seoul, South Korea

The Korean Institute of Chemical Engineers (KIChE) represents the society of Korean Chemical Engineers. The KIChE was founded in 1962 and it has evolved to a non-profit organization having 5,000 members. It consists of one headquater, 9 local chapters, and 13 technical sections. The KIChE has been a main place for exchanging new ideas on technology, building chemical engineering communities, sharing social activities, and enhancing the collaboration between academia and industry. The KIChE aims at leading academic and industrial sectors by disseminating its activities in publishing journal articles and providing national conferences, academic/industrial workshops, and special tutoring sections. Through the educational and research activities, we would like to contribute to providing effective measures to mankind in important matters related to energy, food, and environment. Based on these activities, this talk will provide a comprehensive overview of past, present, and future of the KIChE and prospect collaboration with the AIChE. This year is the 50th anniversary of KIChE. The KIChE hopes to share the idea for coming new paradigms in Chemical Engineering.

(664a) Corning – A Long Tradition of Innovation That Matters

Marc Giroux, Manufacturing Technology & Engineering, Corning Incorporated, Corning, NY

Corning Incorporated is the world leader in specialty glass and ceramics. Drawing on more than 160 years of materials science and process engineering knowledge, Corning creates and makes keystone components and materials that enable high-technology systems for consumer electronics, mobile emissions control, telecommunications and life sciences. The company grows primarily by internal innovation and at the very core of this is a significant industrial R&D lab and engineering center in Corning, NY.

While the connection between glass & ceramics and Chemical Engineering may not at first be obvious, it is very real. This talk will review some of the innovations which have been prominent in our history. It will also cover some current innovations in more detail and highlight the impact Chemical Engineers have on them as well as the reverse: the impact one of them may have on Chemical Engineers.

(665a) Sustainable Design and Operations At the David L. Lawrence Convention Center

Angelica Ciranni, Sports & Exhibition Authority of Pittsburgh and Allegheny County, Pittsburgh, PA

In 2003, the David L. Lawrence Convention Center (DLCC) received a Gold rating under the LEED for New Construction rating system, making it the first convention center to receive LEED certification. At that time, the DLCC was also the world's largest green building at 1.5 million square feet. From the early stages of design through current operations, the DLCC has remained committed to sustainability. In order to demonstrate the sustainable operations of the facility, the DLCC registered under the LEED for Existing Buildings: Operations & Maintenance (O&M) rating system and was certified Platinum in April 2012. The DLCC is the first convention center to receive two LEED ratings.

Concurrent with the LEED for Existing Buildings: O&M documentation process, the DLCC underwent a "building in operation" study that examined building performance, facility management, occupant satisfaction, and organizational sustainability. A thorough set of analyses contributed to the findings, including quantitative and qualitative assessments of energy, natural ventilation, water, site, transportation, waste, purchasing, commissioning, occupant comfort, and indoor air quality.

This session will provide an overview of the case study findings and the LEED for Existing Buildings: O&M process for the host facility of the AIChE Annual Meeting, including specific strategies and lessons learned.

Variations in Temperature in Mini-Roof Structures Employing Different Roofing Materials

Jason T. Kirby[1], Ph.D., Ronald D. Sherrod[2], Matt Winslett[3], and Robert W. Peters[4*], Ph.D., P.E.

[1]Associate Professor of Environmental Engineering
Department of Civil, Construction and Environmental Engineering
University of Alabama at Birmingham
1075 13th Street South
Birmingham, Alabama 35294-4440
Phone: (205)-934-8434; e-mail: rwpeters@uab.edu

[2]Undergraduate Student
Department of Chemical Engineering
Worcester Polytechnic Institute
100 Institute Road
Worcester, Massachusetts 01609-2280

[3]Engineering Manager, Energy Management
Facilities Management Department
University of Alabama at Birmingham
822 Facilities Support Building
Birmingham, Alabama 35294-4564

[4*]Professor of Environmental Engineering
Department of Civil, Construction and Environmental Engineering
University of Alabama at Birmingham
1075 13th Street South
Birmingham, Alabama 35294-4440
Phone: (205)-934-8434; e-mail: rwpeters@uab.edu

ABSTRACT

The overall goal of green roof research at the University of Alabama at Birmingham (UAB) campus is to identify a suitable roofing material combination that reduces energy cooling costs. Since 2008, ambient temperature data has been recorded to better understand how the roof heats and cools on a diurnal, seasonal, and yearly basis. During this study, 15 mini-roof combinations were observed for trends in internal temperatures. Conventional roofing systems included river rock, crushed marble chips, black roofs, reflective (white) roofs, concrete pavers coated with epoxy, and an alternative roofing system involving vegetated roofs (planted with sedum plants). To investigate the thermal properties of the roofing structures, an ambient temperature probe was placed inside of each roof to record data every 10 minutes.

The University of Alabama at Birmingham (UAB) is located in the Southeastern United States. This region is characterized by its hot summers and relatively mild winters. Due to these hot summers during which the temperatures can approach 37.8°C (100°F), along with conditions of high humidity, it is crucial to have a roof which minimizes cooling costs. These hot summers also create an urban heat island effect in the Birmingham area due to the current roofing structures and facility usage. An urban heat island is defined on the basis of a temperature difference between urban and rural area, caused by human modifications to the natural geographic landscape. The human modifications observed here are the buildings and roads added to the natural habitat. These structures retain the heat transferred from the sunlight increasing the temperature. This property results in an increased temperature in the highly modified areas (urban downtown areas) versus a less modified area (rural areas).

Various roofing materials are being studied for their thermal efficiency properties; roofing materials studied include black roofs, white reflective roofs, river rock, crushed marble chips, cement blocks, and vegetated roofs [having sedum plants (*Sedum hispanicum*)]. The temperature behavior both on a daily basis and on an annual basis vary in a sinusoidal fashion. Additionally, a pilot vegetated roof was installed on top of Hulsey Center in July 2008. This pilot green roof occupies ~1388.0 m^2 (14,940 ft^2) and contains ~20,000 sedum type plants. Utility bill information has been collected on natural gas usage, water usage, and electricity usage for ~3 years prior to the installation of the vegetative roof, and for more than a year since installation of the green roof. Reductions in the usage of natural gas, water, and electricity has been observed and documented since installation of the green roof on top of Hulsey Center.

Historically, studies on green roofs have explored their energy performance compared with traditional roofs. Thermal performance indicated a significant reduction (~40%) of a building cooling load during the summer period [Spala *et al.*, 2008]. Similar results were achieved for a nursery school, with reductions ranging from 6% to 49%, and reduction ranging from 12% to 87% on the last floor of the nursery school [Santamouris *et al.*, 2007]. Wong *et al.* [2007] note that green roofs tend to experience lower surface temperatures than the original exposed roof, especially in areas well covered by vegetation. When green roofs are well covered by vegetation, the resulting substrate moisture will tend to keep substrate temperature lower than the original exposed bare roof. These studies determined that over 60% of the heat gain was mitigated by vegetated roof systems. Summertime data have indicated significant lower peak roof surface temperature and higher nighttime surface temperature for green roofs as compared to conventional roofs [Sonne, 2006].

After acquiring the necessary raw data from the mini-roof sensors, we noticed that the temperatures cycled in a sinusoidal fashion on yearly and daily time frames, so using the general form sine function was an appropriate model to describe the data. We determined the frequency and amplitude for all fifteen mini-roof systems. The determined functions supported that most roofs are statistically different from one another from an amplitude aspect but the phase angles are statistically the same. We also have discovered that most roofs are significantly different from the standpoint of the average mean roof temperature, but the significance is most prevalent in the summer months and during the other times of the year the roofs behave in a more similar fashion to one another. The model does well in describing the average daily temperature during the year.

Key Words: Roofing materials, temperatures of roofing structures, thermal properties
*Corresponding author

Introduction/Background

Green building design, as defined by the United States Environmental Protection Agency (USEPA), is "the practice of creating structures and using processes that are environmentally responsible and resource-efficient throughout a building's life-cycle from design, to construction, operation, maintenance, renovation and ultimately deconstruction" [EPA, 2010]. However, according to the United States Green Building Council (USGBC), commercial buildings in the US alone account for:

- 65% of electricity consumption,
- 30% of greenhouse gas emissions,
- 30% of raw materials use,
- 30% of waste output (136 million tons annually), and
- 12% of potable water consumption [USGBC, 2010].

The building infrastructure (Residential, Commercial, Institutional, and Industrial) in the United States (US) consumes over two-thirds of the nation's electricity demand and accounts for one-third of all domestic energy consumption [EPA, 2009]. Regrettably, conventional forms of energy production have an adverse impact on natural ecosystems. Collectively, our buildings contribute to 38.9% of the nation's total greenhouse gas emissions. Faced with rising energy costs, diminishing fuel resources and emerging environmental concerns, scientific research has begun to address these challenges by adopting sustainable or green building alternatives.

Common Roofing Systems

The choice and type of urban roofing system is dependent on environmental concerns such as sun exposure and meteorological factors such as temperature, wind and rain. Some roofs may have covenants to determine their usage. Different systems have been developed and designed to perform at its most effective given the exposure conditions of the building location. As a result, many roofing systems exist and are commonly used by the construction community.

There are many roofing systems used today that depend on the building type, whether is for residential or for commercial applications. The most common systems are listed as follows [Reid, 2000].

- Built-up,
- Preformed metal, and
- Shingles.

A brief description of the three roofing system is provided involving advantages, disadvantages and durability characteristics of each.

Built-Up

Built-up roofing systems are most common for flat roof applications. They consist of several layers of material built-up from the internal structural frame of the building. They are most commonly made of alternating layers of plywood, felt, asphalt, tar and gravel [Reid, 2000]. The major advantage of this system is its waterproofing characteristics, and out of all roofing systems, built-up roofs are considered to have the highest degree of waterproofing. However, the system has poor fire resistance, generally requires a professional to install, and is difficult to locate roof leaks. The system typically lasts for 22.5 years [Cash, 1999; Editors of Time Life Books, 1979].

Preformed Metal

Another common roofing system is made of preformed metal panels. The panels are generally made from aluminum, steel and copper [Reid, 2000]. They are most commonly found with contemporary designs. They come in flat, ridged, ribbed or corrugated forms. These systems are easy to install and repair, and can be painted any color. They are considered fire resistant. The panels are easily damaged by wind, falling trees and tree limbs, or any other type of contact. They generally last for 25 years [Cash, 1999; Editors of Time Life Books, 1979].

Shingles

Shingle roofs come in many forms. Shingles are commonly made of asphalt, slate, wood and clay tiles [Reid, 2000]. Asphalt shingles are the most popular, especially for residential construction. They are available in a variety of sizes, weights, and colors. They require little maintenance and are easy to install. However, they are considered to have poor fire resistive qualities. Asphalt shingles generally last for 16–17 years with proper maintenance [Cash, 1999]. Slate and ceramic tiles are highly expensive due to their aesthetic and durability properties. They are fire resistant and have attractive appearances. Yet, they are extremely heavy and require strong structural support. They are also very brittle, and require increased amount of time to install which often entails specialized tools. Repairs are in most cases difficult. Slate and ceramic tiles are considered the most durable, lasting for 20–100 years with proper maintenance. Wood shingles and shakes are also attractive options. They are easy to install and are a natural insulator. However, they are highly flammable and require treatments for weather and insect protections. They generally last for 10 to 20 years with proper maintenance [Cash, 1999; Editors of Time Life Books, 1979].

Urban Heat–Island Reduction and Building Energy Conservation

For millions of Americans living in and around cities, heat islands are of growing concern. This phenomenon describes urban and suburban temperatures that are 2 to 10°F (1 to 6°C) hotter than nearby rural areas. Elevated temperatures can impact communities by increasing peak energy demand, air conditioning costs, air pollution levels, and heat-related illness and mortality. The Environmental Protection Agency (EPA) recommends installing cool or vegetative green roofs, planting trees and vegetation and switching to cool paving materials as a way of reducing the negative effects of urban heat islands. The EPA says green roofs, if installed widely in a city, can contribute to heat island reduction by replacing heat-absorbing surfaces with plants, shrubs, and small trees. The vegetation cools the air through evapotranspiration (or evaporation of water from leaves). Planted rooftops remain significantly cooler than a rooftop constructed from traditional heat-absorbing materials. Further, green roofs reduce summertime air conditioning demand by lowering heat gain to the building.

Energy Modeling

Energy modeling (i.e., energy simulation) is a method for predicting the energy consumption of an occupied structure. Building energy analysis must consider numerous thermal characteristics including: wall and roof materials, the size and orientation of the building, how the building is occupied and operated, as well as influences from the local climate.

The surface temperature of a roof exposed to solar radiation, the resulting heat flow into the building, along with associated indoor temperatures and cooling needs depend on the effect of solar radiation, surface absorptivity, ambient air temperature and wind speeds adjacent to the surface [Givoni, 1994]. When vegetative roofs are considered, because of added thermal mass, it is also important to take into account hourly heat transfer when determining energy consumption; as the heat flux through a vegetated roof can be quite different from conventional roofing.

R-values and *U*-values have been used for many years as a measurement of a building envelope's thermal performance. However, these attributes do not fully take into account the effects of thermal mass, and by themselves, are inadequate in describing the heat transfer properties of construction assemblies with significant amounts of thermal mass [Kosny *et al.*, 2007]. Vegetated roofs are more dependent on the interaction between the roofing systems' unit weight, density, thermal conductivity, moisture content, vegetal coverage and specific heat. Therefore, it is often necessary to utilize computer software, which incorporates these elements into the analysis of high thermal mass roof structures and associated energy consumption. The steady-state *R*-values traditionally used to measure energy performance will not accurately capture the complex, dynamic thermal behavior of vegetated roof systems.

Vegetated (Green) Roof Systems

Green roofs are engineered ecosystems that rely on vegetation to provide benefits such as reduction of roof temperatures and stormwater retention [Wolf and Lundholm, 2008]. Green roofs offer benefits of reducing stormwater runoff, improving air and water quality, and providing habitat and biodiversity for urban centers [Wong *et al.*, 2003][0]. Hydrologic modeling has demonstrated that widespread green roof implementation can significantly reduce peak runoff rates, particularly for small storm events [Carter and Jackson, 2007]. By combining the Green-Ampt method with evapotranspiration of green roofs, Roehr and Kong [2010] estimated the potential runoff reduction achieved by green roofs is 20%. Green roofs provide an excellent option to improve stormwater runoff [Hathaway *et al.*, 2008]. Green roofs are primarily valued based on their increased roof longevity, reduced stormwater runoff, and decreased building energy consumption [Clark *et al.*, 2008]. Carter and Jackson [2007] noted that research studies have primarily been focused on roof-scale processes such as individual roof stormwater retention, plant growth, or growing medium composition. Few studies have examined the impact that widespread green roof application could have on the hydrology of a real-world watershed [Carter and Jackson, 2007]. A major barrier to increasing the prevalence of green roofs is the lack of scientific data available to evaluate their applicability to local conditions [Williams *et al.*, 2010].

Green roofs are typically classified as being either an intensive or extensive roof [Green Roof Plan, undated]. Intensive green roofs are often used on commercial buildings in order to have large green areas that incorporate all sizes and types of plants. These roofs use grasses, ground covers, flowers, shrubs and even trees. They often include paths and walkways that travel between different architectural features to provide space where people can interact with the natural surroundings. Intensive green roofs, sometimes termed "rooftop gardens", utilize planting mediums that have greater depth than extensive green roofs; the deeper soil allows intensive roofs to accommodate large plants and various plant groupings. Intensive green roofs require more maintenance than extensive green roofs because of the plant varieties they will support.

Extensive green roofs have a planting medium that ranges from 1.6 to 6 inches deep. Typically, drought-tolerant sedums (succulent plants) and grasses are used since they are shallow-rooted and use little water. Plant diversity on these roofs is kept low to simplify care and to be sure all plants have similar moisture requirements.

Extensive green roofs can significantly reduce both the timing and magnitude of stormwater runoff relative to a typical impervious roof [Schroll *et al.*, 2011]. They note, however, that regional climatic conditions such as seasonality in rainfall and potential evapotranspiration can strongly alter the stormwater performance of vegetated roofs. Factors such as type of green roof and its geometrical properties (slope), soil moisture characteristics, season, weather and rainfall characteristics, age of the vegetated roof, and vegetation affect the runoff dynamics from green roofs [Berndtsson, 2010]. Fioretti *et al.* [2010] noted that green roofs significantly mitigate storm water runoff generation, as well reducing the daily energy demand. Aitkenhead-Peterson *et al.* [2011] note that most studies on runoff quality from green roofs have been conducted in cooler northern climates. Villarreal and Bengtsson [2004] recommended the use of a combination of best management practices; additionally, they observed that green roofs are effective at lowering the total runoff from Augestenborg (Sweden) and that detention ponds should successfully attenuate storm peal flows. Niu *et al.* [2010] noted that over the lifetime of a green roof (~40 years), the net present value is ~30% to 40% less for a green roofs as compared with conventional roofs (not including green roof maintenance costs). Kirby *et al.* [2011] note that extensive vegetated roof systems offer at least 1 16% enhancement in reducing stormwater runoff as compared to conventional roofs. Clark *et al.* [2008] further note that the additional upfront investment of a green roof is recovered at the time when a conventional roof would be replaced. Rosatto *et al.* [2010] concluded that green roofs contribute positively in reducing runoff, with greater retention with vegetated plots and thicker substrate.

Vegetated roof systems have a number of advantages over that of conventional roof systems. Benefits associated with green roof systems include [Green Roofs for Healthy Cities, undated]:

- Urban greening has long been promoted as an easy and effective strategy for beautifying the built environment and increasing investment opportunity.
- With green roofs, water is stored by the substrate and then taken up by the plants from where it is returned to the atmosphere through transpiration and evaporation.
- Depending on the plants and depth of growing medium, during the summer, green roofs retain 70% to 90% of the precipitation that falls on them; in winter they retain between 25% to 40%.

- Green roofs not only retain stormwater, but also moderate the temperature of the water and act as natural filters for any of the water that runs off.
- Green roofs reduce the amount of stormwater runoff and delay the time at which runoff occurs, resulting in decreased stress on sewer systems at peak flow periods.
- Through the daily dew and evaporation cycle, plants on vertical and horizontal surfaces can cool cities during hot summer months and reduce the Urban Heat Island (UHI) effect. The UHI is also mitigated by the covering some of the hottest surfaces in the urban environment, such as black rooftops.
- Green roofs can also help reduce the amount of dust and particulate matter throughout the city, as well as the production of smog. This plays a role in reducing greenhouse gas emissions and adapting urban areas to a future climate with warmer summers.
- Green roofs help to achieve the principles of smart growth and positively affect the urban environment by increasing amenity and green space.
- The greater insulation offered by green roofs can reduce the amount of energy needed to moderate the temperature of a building, as roofs provide the greatest heat loss in the winter and the hottest temperatures in the summer.
- The presence of a green roof decreases the exposure of waterproofing membranes to large temperature fluctuations, which can cause micro-tearing, and ultraviolet radiation.
- Green roofs have excellent noise attenuation, especially for low frequency sounds. An extensive green roof can reduce sound from outside by 40 decibels, while an intensive one can reduce sound by 46-50 decibels.
- Green roofs can sustain a variety of plants and invertebrates, and provide a habitat for various bird species.

Historically, studies on green roofs have explored their energy performance compared with traditional roofs. Thermal performance indicated a significant reduction (~40%) of a building cooling load during the summer period [Spala *et al.*, 2008]. Similar results were achieved for a nursery school, with reductions ranging from 6% to 49%, and reduction ranging from 12% to 87% on the last floor of the nursery school [Santamouris *et al.*, 2007]. Wong *et al.* [2007] note that green roofs tend to experience lower surface temperatures than the original exposed roof, especially in areas well covered by vegetation. When green roofs are well covered by vegetation, the resulting substrate moisture will tend to keep substrate temperature lower than the original exposed bare roof. These studies determined that over 60% of the heat gain was mitigated by vegetated roof systems. Summertime data have indicated significant lower peak roof surface temperature and higher nighttime surface temperature for green roofs as compared to conventional roofs [Sonne, 2006]. The maximum average daily temperature seen for the conventional roof surface was 54.4°C (129.9°F) in his study, while the maximum average day green roof surface temperature was 32.8°C (~21.7°C lower than the conventional roof). Green roofs offer cooling potential (~3.02 kWh/day) to maintain an average room air temperature of 25.7°C (78.3°F) [Kumar, 2005]. Green roofs help minimize environmental burdens, conserve energy, and extend the life span of the roofing system in overall sustainability [Cavanaugh, 2008]. Up to 30% of total rooftop cooling is due to plant transpiration [Takakura *et al.*, 2000]. Bell and Spolek [2009] compared different types of plants for use in increasing the thermal resistance (*R*-value) of green roofs, and found that ryegrass delivered the highest effective *R*-value compared with bare soil, *Vinca major, Trifolium repens,* and *Sedum hispanicum.* Also, though increasing the depth of bare soil from 5 to 14 cm (2.0 to 5.5 inches) increased the *R*-value, no difference was found for different depths of planted soil. This implies that the bulk of benefit toward *R*-value is from evapotranspiration and leaf shading, rather than the moist soil [Bell and Spolek, 2009].

There are several detailed building simulation programs (BSPs) that take into consideration the complete interaction between all thermal-based elements. The most popular BSPs are A Simplified Energy Analysis Method (ASEAM), Building Design Advisor (BDA), Building Load Analysis and

Systems Thermodynamics (BLAST), Builder Guide, Bus^{++}, Dynamic Energy Response of Buildings (DEROB), DOE-2, Energy-10, Energy Plus, ENERPASS, ENER-Win, ESP, FEDs, Home Energy Saver, Hot 2000, TRNSYS, and VisualDOE [Selim, 2003; Selim, 2008; Mansour, 2001].

UAB has utilized Visual DOE in the past with great success in the analysis of innovative structures designed for energy efficiency. VisualDOE uses the DOE 2 calculating core and provides output in both numerical and graphical forms. This software is a preferred calculation method due to its cost, previous verification/validation success, ease of use, database support and reasonable input/output requirements. We envision that this computer simulation tool will be able to effective capture the differences in roof types being explored in the purposed research.

Results and Discussion

Thermal Performance of Mini-Roof Structures

Mini-Roofs

During this study, 15 mini-roof combinations were observed for trends in internal temperatures. The various 15 mini-roof combinations are summarized in Table 1. Several of the mini-roof structures are depicted in Figure 1. This photo shows the layout of the 15 mini-roofs, and a vegetated roof from which surface temperatures of the mini-roofs were measured periodically using an infrared thermometer (see Figure 1).

(a) (b)

Figure 1. (a) Layout of the 15 mini-roods;
(b) vegetated mini-roof (surface temperatures were measured using an IR thermometer.

The roofing materials used are all standard commercial flat roof materials. Flat roof materials were only looked at during the study, since the primary application for the roofing combinations will be on a commercial flat roof top, and not a slanted roof structure. Each mini-roof is 2.4-m (8.0-ft) long x 1.2-m (4.0-ft) wide x 1.2-m (4.0-ft) deep (see Figure 1). A number of different roofing systems are being examined for their energy performance. All roofs are insulated with 5.1 cm (2.0-in) of extruded polystyrene. Then the particular roofing combination being investigated is applied over the insulation and sealed. The roofs also include a proper drainage spout, to ensure correct water evacuation, such as on a real roof.

The roofing systems being studied using the 15 mini-roofs are listed in Table 1.

Table 1. Mini-roof descriptions.

Mini-Roof No.	Mini-Roof Description
1 (Sensor B)	White TPO/PVC/ Elvaloy fully adhered, FiberTite Membrane.
2 (Sensor C)	Black 60-mil EPDM fully adhered/coated/white urethane, Mule Hide Membrane.
3 (Sensor D)	Black 60-mil EPDM fully adhered, Mule Hide Membrane.
4 (Sensor E)	Beige TPO/PVC/Elvaloy fully adhered, FiberTite Membrane.
5 (Sensor F)	White granular modified, Firestone SBS Modified Membrane.
6 (Sensor G)	Black granular modified, Firestone SBS Modified Membrane.
7 (Sensor H)	Black granular modified coated/white urethane, Firestone SBS Modified Membrane.
8 (Sensor I)	Bituthene IRMA with lightweight "T Clear" pavers.
9 (Sensor J)	Bituthene IRMA with river rock ballast.
10 (Sensor K)	Bituthene IRMA with vegetative green roof, ½-in. drain mat, 350-lbs dry soil.
11 (Sensor L)	Bituthene IRMA with vegetative green roof, 1-in. drain mat, 350-lbs dry soil.
12 (Sensor M)	Black 60-mil EPDM loose, ballasted with river rock, Mule Hide Membrane.
13 (Sensor N)	Black 60-mil EPDM loose, ballasted with #300 marble chips, Mule Hide Membrane.
14 (Sensor O)	White TPO/PVC/Elvaroy loose laid, ballasted with river rock, FiberTite Membrane.
15 (Sensor P)	Bituthene IRMA with vegetative green roof, ½-in. drain mat, 350-lbs dry soil.
Sensor S	Sensor inside mini-roof No.10 (inside the soil of the green roof).
Sensor T	Sensor under the white TPO/PVC/Elvaloy loose laid, ballasted with river rock.

To investigate the thermal properties of the roofing structures an ambient temperature probe was placed inside of each roof (see Figures 2 and 3) recording temperature data every 10 minutes of each day, for more than 3 years. This data was then automatically sent to a data logger and placed into an Excel file for review later. The temperature probe reports the temperature to the nearest hundredth of a degree Centigrade.

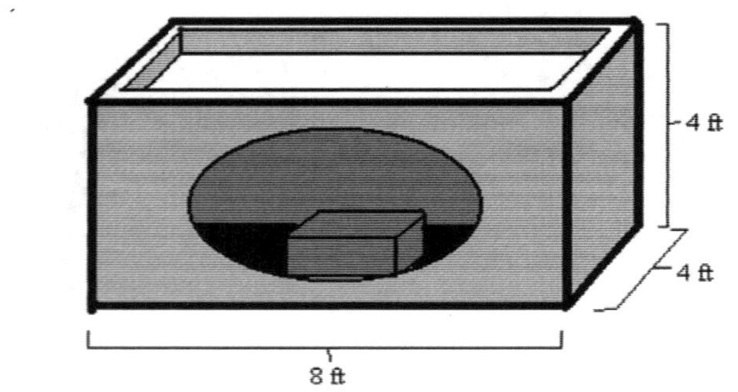

Figure 2. Depiction of a Typical Mini-Roof System.

Figure 3. Mini-roof system showing temperature sensor installed inside a mini-roof.

Figure 4 presents some typical temperature profiles on several different days.

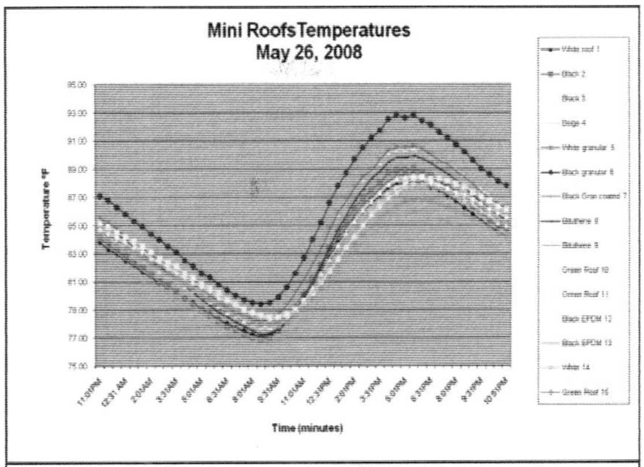

Figure 4a. Temperature profile inside the various mini-roofs on May 26, 2008.

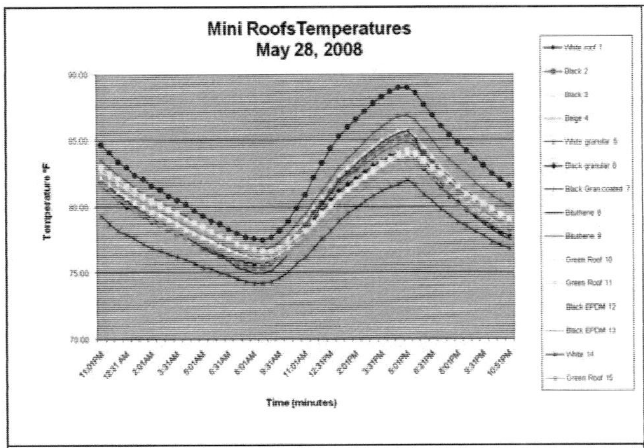

Figure 4b. Temperature profile inside the various mini-roofs on May 28, 2008[Nunez and Peters, 2008].

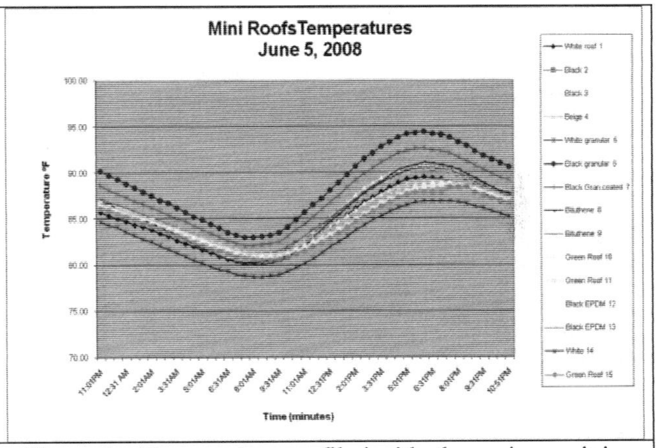

Figure 4c. Temperature profile inside the various mini-roofs on June 5, 2008.

The results from these mini-roof structures have shown the following trends [Nunez and Peters, 2008]:

- Clean white roofs resulted in consistently lower temperatures inside the mini roof than the other roofing materials.
- Black roofs resulted in the highest temperature readings.
- Green roofs resulted in temperatures typically ~1.1-1.7°C (2-3°F) higher than the white roofs; however, they will dampen the drainage of rainfall during a rain storm through the retention of water onto the soil.
- Bituthane (river rock) performs only slightly better than black roofing materials.
- White granular roofing behaved similarly to black granular roofing materials.
- Over time, the reflective (white) roofs become dirty, losing some of their reflectivity, resulting in the roof being less energy efficient.
- White marble chips behaved slightly worse than green roofing materials, but considerably better than black roofing and granular roofing materials.
- While it is too early to come to a definitive conclusion, preliminary evaluations indicate that the "white" and "green" roofs both significantly reduce the roofs surface temperature and therefore, the air temperatures above and around the roof.

Surface temperature measurements during the months of June and July on the various roofing materials used with our mini-roof systems were collected. During this time period, roofing surface temperatures ranging from 20.6°F to 82.2°C were observed. The lower surface temperature values, 20.6°C to 48.3°C, are found in a loose rock or stone roof combination. The higher temperature values occurred on the roofing combinations primarily made of a coating or membrane; these roofs exhibited temperatures roughly between 25.6°C to 82.2°C. The vegetative mini-roofs exhibited surface temperatures ranging between 21.7°C to 52.2°C, while their large counterpart, the pilot roof on top of Hulsey Center, exhibited higher temperatures ranging from 31.7°C to 61.1°C. When looking at the specific roofs contained in each subset of Table 1, it was observed that generally, the lighter colorization of the roof resulted in cooler temperatures. For example, the white Firestone SBS is cooler than the black Firestone SBS on any given day. This thermal property is observed due to the reflectivity of the roof. The darker colored roofs absorb more incoming light radiation than the light colored roofs causing the dark roofs to become hotter. (Since the roofs temperatures observed are taken during the late spring and early summer, it can be inferred that overall roof temperature will increase during late summer).

Photographs of several of these mini-roofs are presented in Figures 5 through 9. Figure 10 presents a typical temperature profile of the 15 mini-roofs during the course of a typical summer week. Series 1 through 5 denoted in the figure refer to the fifteen mini-roofs listed in Table 1. This figure shows a cyclical nature of the temperature readings over each day, generally showing a sinusoidal behavior of temperature; the temperature is cool in the morning, warms up, and is at its hottest during mid-afternoon, and then cools down during evening hours.

Trends seen in statistical comparisons of internal temperatures of similar roofs (using the null hypotheses: $\mu_1 \geq \mu_2$) are summarized below [Peters *et al.*, 2011]:

- Between the two river rock roofs, Roof 9 will most likely always be hotter or equal to roof 12.
- Between the 3 vegetative roofs, all the roofs are statistically equal to each other in thermal properties.
- Both SBS Firestone roofs are statistically the same, but roof 6 is usually hotter.
- The TPO/PVC/Elvaloy Roofs are statistically the same.
- The 60-mil EPDM roofs are also statistically the same.

Figure 5. Black mini-roof.

Figure 6. White (reflective) mini-roof.

Figure 7. Mini-roof equipped with river rocks.

Figure 8. Mini-roof equipped with crushed marble.

Figure 9. Vegetative mini-roof equipped with *Sedum* plants.

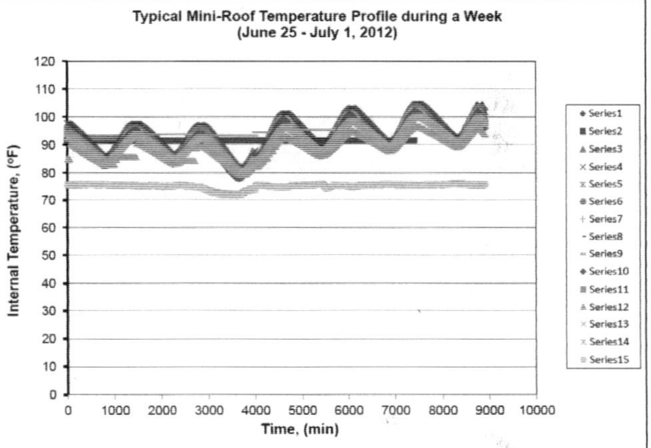

Figure 10. Typical internal mini-roof temperatures during the course of a week.

After acquiring the necessary raw data from the mini-roof sensors, Microsoft Excel (version 2010) was used to plot the data to discover trends in the temperature readings. Subsequent to discovering the trend, a mathematical model was fit to the data. It was noticed that the temperatures cycled in a sinusoidal fashion on yearly and daily time frames, and therefore a general form sine function was utilized as a potential model. Fourier transforms were utilized in order to determine the oscillation frequency of the temperature (ω_0 from the general form $x = A \sin(\omega_0 t + \phi) + C$) by transforming the time domain of the collected raw data into a frequency domain. Two major peaks were discovered from the spectrum: one representing the yearly frequency and the other representing the daily oscillations. After calculating the frequencies, the phase angle (ϕ) and the amplitude (A) of the general form were determined through a regression analysis.

The modeling procedure was applied for all 15 mini-roofs in the study. The developed sine-wave functions indicated that most roofs were statistically different from one another from an amplitude aspect but the phase angles were statistically the same. It was also discovered that almost all roofs had significantly different average mean roof temperatures, but the significance was mostly prevalent in the summer months. During other times of the year, the roofs behaved in a similar fashion (see Tables 2 and 3 for statistics and means of roofs). The fitted sine wave functionalities for the 15 mini-roofs are listed in Table 4.

Table 2.	**Summary of statistics for the phase angles from curve fitting.**
Statistics of Phase Angles	
Mean	4.0303
Standard Error	0.00343
Median	4.0276
Mode	4.0242
Standard Deviation	0.01329
Sample Variance	0.000177
Kurtosis	12.0174
Skewness	3.3228
Range	0.05512
Minimum	4.0211
Maximum	4.0762
Sum	60.4541
Number (count)	15

Table 3.	**Summary of statistics for the means from curve fitting.**
Statistics of Means	
Mean	66.67
Stand Error	0.18
Median	66.62
Mode	N/A
Standard Deviation	0.68
Sample Variance	0.46
Kurtosis	3.11
Skewness	1.11
Range	3.06
Minimum	65.42
Maximum	68.47
Sum	1000.12
Number (count)	15

Table 4. Fitted sine-wave functionalities describing the internal temperatures in the mini-roofs.

Roof	**Final Equation**	**Amplitude from sinefind.exe**	**Phase Angle**	r^2	**Mean**
1	$T = 20.1632*\sin((2*\pi*3.17894e(-8)*t)+4.021975)+66.68283$	20.163	4.022	0.696	66.683
2	$T = 20.5499*\sin((2*\pi*3.18e(-8)*t)+4.02272)+66.63564$	20.550	4.023	0.701	66.636
3	$T = 20.392*\sin((2*\pi*3.18e(-8)*t)+4.07624)+66.47947$	20.392	4.076	0.712	66.479
4	$T = 20.482*\sin((2*\pi*3.18e(-8)*t)+4.021125)+66.35505$	20.482	4.021	0.698	66.355
5	$T = 20.4891*\sin((2*\pi*3.18e(-8)*t)+4.03214)+67.54093$	20.349	4.032	0.701	67.541
6	$T = 20.7812*\sin((2*\pi*3.18e(-8)*t)+4.03142)+68.46883$	20.781	4.031	0.699	68.469
7	$T = 20.2886*\sin((2*\pi*3.18e(-8)*t)+4.02415)+67.08382$	20.289	4.024	0.701	67.084
8	$T = 20.7812*\sin((2*\pi*3.18e(-8)*t)+4.02415)+66.18969$	20.781	4.024	0.695	66.190
9	$T = 20.3285*\sin((2*\pi*3.18e(-8)*t)+4.024975)+66.91944$	20.329	4.025	0.701	66.919
10	$T = 19.8608*\sin((2*\pi*3.18e(-8)*t)+4.02758)+66.6981$	19.861	4.028	0.697	66.698
11	$T = 20.0128*\sin((2*\pi*3.18e(-8)*t)+4.026215)+66.62328$	20.013	4.026	0.697	66.623
12	$T = 20.0479*\sin((2*\pi*3.18e(-8)*t)+4.027853)+66.27383$	20.048	4.028	0.694	66.274

13	T = 19.9914*sin ((2*π*3.18e(-8)*t) +4.0325)+66.47272	19.991	4.033	0.696	66.473
14	T = 20.0661*sin ((2*π*3.18e(-8)*t) +4.032494)+65.42343	20.066	4.032	0.689	65.423
15	T = 19.4628*sin ((2*π*3.18e(-8)*t) +4.028672)+66.27177	19.463	4.029	0.692	66.272

These results of this study were in agreement with research conducted by Watson [2011] addressing the urban heat island effect for the City of Birmingham by studying the effect from different building facades, building materials and seasonal traits. The data indicates amplitudes (*A*) ranging from 19.5 to 20.7 degrees and phases angles roughly 4.2 to 4.3 radians. The r^2 values ranged from 0.69 to 0.71showing the data modeling was applicable to understanding the trends. The best model developed was for roof 3 (see Figure 11), having the highest correlation coefficient of all 15 mini-roofs.

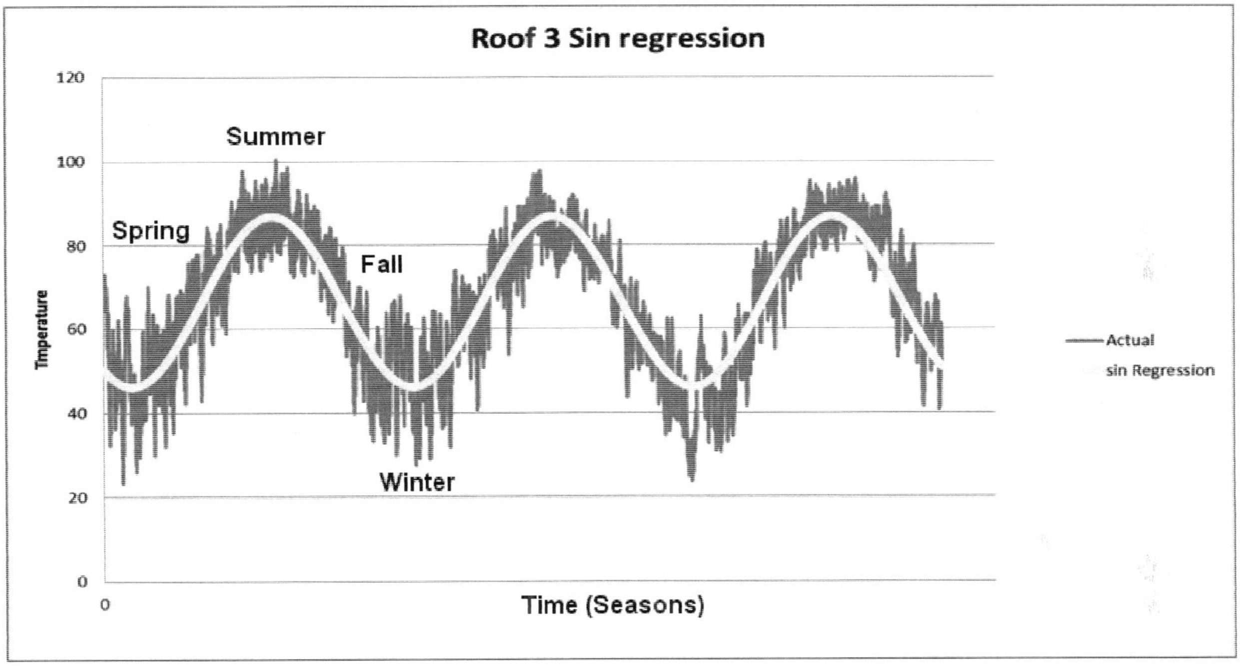

Figure 11. Fitted sine wave regression for a mini-roof system.

Based on the information collected on these mini-roof systems, the UAB Facilities Management Department decided to install an extensive vegetated roof on top of Hulsey Center as a roofing retrofit. UAB wants to obtain more fundamental information and knowledge for establishing green roofs in the southeastern U.S. In the Birmingham, Alabama area, fairly high rainfalls [approximately 132.1 cm/year (52 inches/year) on average] are obtained. However, during the summer months, it is common to have periods of drought with minimal rainfall and very hot and humid days [with temperatures in the 32.2^{+o}C (90+°F range]. Such climatic conditions require plants that can withstand both significant rainfall events and drought conditions. The current pilot green roof on top of Hulsey Center was installed in July 2008. Research conducted at Penn State University shows that green roofs planted with sedum plants reduce the building energy costs [Penn State Center for Green Roof Research, undated (a, b)]. The Facilities Management Department at UAB maintains records of the utilities bills (e.g., natural gas, water, and electricity) for each building on campus. The quantities and costs of utilities prior to (5 years) and after implementation (~3 years) of the pilot green roof have resulted in building energy reductions of 20% to 25%.

Summary and Conclusions

The behavior of different roofing materials affects the heating loads placed upon the buildings and their heating, ventilation, and air conditioning (HVAC) systems. Black roofs resulted in the highest temperature readings. Black roofing materials and bituthane (river rock) perform the poorest of the roofing materials tested, resulting in the highest heating load being placed on the building infrastructure. White granular roofing behaved similarly to black granular roofing materials. Clean white roofs resulted in consistently lower temperatures inside the mini-roof than the other roofing materials; however, over time, the reflective (white) roofs become dirty, losing some of their reflectivity, resulting in the roof being less energy efficient. Green (vegetated) roofs are fairly efficient in terms of energy performance due to evapotranspiration effects. Green roofs resulted in temperatures typically ~1.1-1.7°C (2-3°F) higher than the white roofs; however, they will dampen the drainage of rainfall during a rain storm through the retention of water onto the soil, and thereby lessen the discharge into stormdrains. Green roofs can significantly reduce stormwater runoff, reduce peak flow quantities, and lengthen the time of concentration from roofing structure [Kirby *et al.*, 2011]. "White" and "green" roofs both significantly reduce the roofs surface temperature and therefore, the air temperatures above and around the roof.

Hypothesis testing indicated that, between the 3 vegetative mini-roofs, all the min-roofs are statistically equal to each other in thermal properties. Both SBS Firestone mini-roofs are statistically the same, but mini-roof 6 is usually hotter. The TPO/PVC/Elvaloy mini-roofs are statistically the same. The 60-mil EPDM mini-roofs are also statistically the same.

The temperature varies in a sinusoidal fashion both during the course of the day and on an annual basis. For the various mini-roof structures, the phase angle (ϕ) and the amplitude (A) of the general form were determined through a regression analysis. The developed sine-wave functions indicated that most roofs were statistically different from one another from an amplitude aspect but the phase angles were statistically the same. It was also observed that almost all roofs had significantly different average mean roof temperatures, but the significance was mostly prevalent in the summer months. During other times of the year, the roofs behaved in a similar fashion.

Based on the information collected on these mini-roof systems, the UAB Facilities Management Department installed a vegetated roof on top of Hulsey Center as a roofing retrofit. The vegetated roof is ~1388.0 m^2 (14,940 ft^2) in area, and contains approximately 20,000 sedum plants. Utility bill information both prior to and after implementation of the green roof were gathered for electricity, natural gas, and chilled water were collected. The costs of utilities prior to and after implementation of the pilot green roof indicated utility bill (energy) savings of ~20% to 25% (compared to the case prior to implementation of the vegetated roof system).

Acknowledgments

The financial support and technical support of the UAB's Facilities Management Department is appreciated, including the fabrication and installation of the mini-roofs (on top of the Chemistry Building) and pilot roof on top of Hulsey Center. The input and technical assistance provided by Matt Winslett and Kelly Winnett of the Facilities Management Department in supplying the utility bill information is gratefully acknowledged.

References Cited:

Aitkenhead-Peterson, J.A., B.D. Dvorak, A. Voider, and N.C. Stanley, 2011. "Chemistry of Growth Medium and Leachate from Green Roof Systems in South-Central Texas", ***Urban Ecosystems***, ***14***(1): 17–33.

Bell, H., and G. Spolek, 2009. "Measured Energy Performance of Greenroofs", in ***Proc. Seventh Annual Greening Rooftops for Sustainable Communities Conference***, The Cardinal Group, Toronto, Canada.

Berndtsson, J.C., 2010. "Green Roof Performance towards Management of Runoff Water Quantity and Quality: A Review", ***Ecol. Engrg.***, ***36*** (4): 351–360.

Carter, T., and C.R. Jackson, 2007. "Vegetated Roofs for Stormwater Management at Multiple Spatial Scales", ***Landscape and Urban Planning***, ***80*** (1–2): 84–94.

Cash, C.G., 1999. "The Relative Durability of Low-Slope Roofing", pp. 119–124 in ***Proceedings of the Fourth International Symposium on Roofing Technology***, Gaithersburg, MD, (September 17–19).

Cavanaugh, L.M., 2008. "Redefining the Green Roof", ***J. Archit. Eng.***, ***14*** (1): 4–6.

Clark, C., P. Adriaens, and F.B. Talbot, 2008. "Green Roof Valuation: A Probabilistic Economic Analysis of Environmental Benefits", ***Environ. Sci. Technol.***, ***42***(6): 2155–2161.

Editors of Time Life Books, 1979. ***Roofs and Siding***, Time-Life Books, Alexandria, VA.

Environmental Protection Agency (EPA), 2010. ***Environmental Protection Agency Green Building Design***. www.epa.gov/greenbuilding/.

Environmental Protection Agency (EPA), 2009. ***Buildings and their Impact on the Environment: A Statistical Summary***. www.epa.gov/greenbuilding/pubs/gbstats.pdf.

Fioretti, R., A. Palla, L.G. Lanza, and P. Principi, 2010. "Green Roof Energy and Water Related Performance in the Mediterranean Climate", ***Build. and Environ.***, ***45*** (8): 1890–1904.

Givoni, B., 1994. ***Passive & Low Energy Cooling of Buildings***, John Wiley & Sons, Inc., New York, NY.

Green Roofs for Healthy Cities, undated. Website: http://www.greenroofs.org/index.php/about-green-roofs/2577-aboutgrnroofs/, accessed August 3, 2012.

Green Roof Plan, undated. Website: http://www.greenroofplan.com/intensive-vs-extensive-green-roofs/, accessed August 3, 2012.

Hathaway, A.M., W.F. Hunt, and G.D. Jennings, 2008. "A Field Study of Green Roof Hydrologic and Water Quality Performance", ***Transactions Am. Soc. Agric. & Biolog. Engineers***, ***51***(1): 37–44.

Kirby, J.T., R. Peters, S. Watts, and M. Winslett, 2011. "Experimental Investigation of Runoff Mitigation from Vegetated Roofs", Final report submitted to the Alabama Water Resources Research Institute, Auburn, AL.

Kosny, J., D. Yarbrough, P. Childs, and S. Mohiuddin, 2007. "How the Same Wall Can Have Several Different *R*-Values: Relations between Amount of Framing and Overall Thermal Performance in Wood And Steel-Framed Walls", *Oak Ridge National Laboratory*.

Kumar, R., 2005. "Performance Evaluation of Green Roof and Shading for Thermal Protection of Buildings", ***Build. Environ.***, ***40*** (11): 1505–1511.

Mansour, H.M.A., 2001. "Sustainable Housing Development with Special Reference to Environmental Aspects in Low Cost Housing in Egypt". M.S. Thesis, Cairo University, Cairo, Egypt. 117 pages.

Niu, H., C. Clark, J. Zhou, and P. Adriaens, 2010. "Scaling of Economic Benefits from Green Roof Implementation in Washington, D.C.", ***Environ. Sci. Technol.***, ***44*** (11): 4302–4308.

Nunez, S., and R.W. Peters, 2008. "Effect of Roofing Materials on Building Energy Efficiency", Paper included in the conference proceedings of the 2008 Annual AIChE Meeting, Philadelphia, PA, (November 16–21).

Penn State Center for Green Roof Research, undated (a). Pennsylvania State University website: http://horticulture.psu.edu/cms/greenroofcenter/about_ctr.html; accessed July 5, 2009.

Penn State Center for Green Roof Research, undated (b). "Utility Bills Getting You Down? Plant Stuff Up on the Roof", http://aginfo.psu.edu/news/2000/9/roof.html; accessed July 5, 2009.

Peters, R.W., J.G. Price, R.D. Sherrod, S.A. Watts, J.M. Gohlke, J.T. Kirby, and M. Winslett, 2011. "Roofing Types and Their Effects on the Thermal Properties of Runoff", ***Proc. Alabama Water Resources Conference***, Orange Beach, AL, (September 8–9).

Reid, R.N., 2000. ***Roofing and Cladding Systems: A Guide for Facility Managers***, Fairmont Press, Lilburn, GA.

Roehr, D., and Y. Kong, 2010. "Stormwater Runoff Reduction Achieved by Green Roofs: Comparing SWMM Method to TR-55 Method", pages 1012–1021 in ***Low Impact Development 2010: Redefining Water in the City, Proceedings of the 2010 International Low Impact Development Conference***, San Francisco, CA, (April 11–14).

Rosatto, H.G., D. Laureda, D. Perez, D. Nerrera, M. Meyer, P. Gamboa, G. Villalba, M. Friedrich, M. Bargieta, L.R. Plaza, G. Calvo, M. Miranda, M. Inigo, and E. Quaintenne, 2010. "Water Retention Efficiency of Green Roof Systems", ***Revista de la Facultad de Ciencias Agarias***, ***42*** (1): 213–219.

Santamouris, M., C. Pavlou, P. Doukas, G. Mihalakakou, A. Synneta, A. Hatzibiros, and P. Patargias, 2007. "Investigating and Analysing the Energy and Environmental Performance of an Experimental Green Roof System Installed in a Nursery School Building in Athens, Greece", ***Energy***, ***32*** (9): 178–1788.

Schroll, E., J. Lambrinos, T. Righetti, and D. Sandrock, 2011. "The Role of Vegetation in Regulating Stormwater Runoff from Green Roofs in a Winter Rainfall Climate", ***Ecol. Engrg.***, ***37*** (4): 595–600.

Selim, M.A., 2003. "The Effect of Building Envelope Technology on Energy Conservation in Buildings", M.S. Thesis. Cairo University. Cairo, Egypt.

Selim, R., 2008. "Investigation of Thermal Performance, Indoor Air Quality and Energy Consumption within Tuskegee Healthy House Demonstrative Model", Ph.D. Dissertation,. University of Alabama at Birmingham, Birmingham, AL.

Sonne, J., 2006. "Evaluating Green Roof Energy Performance", ***ASHRAE J.***, ***48*** (2): 59–61.

Spala, A., H.S. Bagiorgas, M.N. Assimakopoulos, J. Kalavrouziotis, D. Matthopoulos, and G. Mihalakakou, 2008. "On the Green Roof System. Selection, State of the Art and Energy Potential Investigation of a System Installed in an Office Building in Athens, Greece", ***Renew. Energy***, ***33*** (1): 173–177.

Takakura, T., S. Kitade, and E. Goto, 2000. "Cooling Effect of Greenery Cover over a Building", ***Energy Build.***, ***31***: 1–6.

United States Green Building Council (USGBC), 2010. United States Green Building Council. www.usgbc.org.

Villarreal, E.L., and A.S.D.L. Bengtsson, 2004. "Inner City Stormwater Control Using a Combination of Best Management Practices", ***Ecol. Engrg.***, ***22*** (4–5): 279–298.

Watson, C.J., 2011. "Sustainable Urban Design Practices that Mitigate Urban Heat Islands and Reduce Energy Consumption", Ph.D. Dissertation, University of Alabama at Birmingham, Birmingham, AL.

Williams N.S.G., J.P. Nicholas, and K.J. Raynor, 2010. "Green Roofs for a Wide Brown Land: Opportunities and barriers for Rooftop Greening in Australia", ***Urban Forestry & Urban Greening***, ***9***(3): 245–251.

Wolf, D., and J.T. Lundholm, 2008. "Water Uptake in Green Roof Microcosms: Effects of Plant Species and water Availability", ***Ecol. Engrg.***, ***33*** (2): 179–186.

Wong, N.H., P.Y. Tan, and C. Yu, 2007. "Study of Thermal Performance of Extensive Rooftop Greenery Systems in the Tropical Climate", ***Build. Environ.***, ***42*** (1): 25–54.

Wong, N.H., S.F. Tay, R. Wong, C.L. Ong, and A. Sia, 2003. "Life Cycle Cost Analysis of Rooftop Gardens in Singapore", ***Building and Environment***, ***38*** (3): 499–509.

(665c) Low-Cost Phase Change Material for Building Envelopes

Ramin Abhari, Syntroleum, Tulsa, OK and **Laura Ford**, University of Tulsa, Tulsa, OK

Studies have shown that incorporating phase change material (PCM) in building envelopes—attic insulation, wall boards, etc.—can reduce energy consumption for home heating and cooling by more than 30%. However, high manufacturing costs have so far limited PCM use to novelty products. This work has developed a low-cost PCM through a selective process for n-paraffin production and continuous high-throughput equipment for its encapsulation. The n-paraffins are made sustainably via hydrodeoxygenation of bio-oils. The heat of fusion and melting point of our n-paraffins made from bio-oils compare well with those of traditional n-paraffins. Heat cycling tests at 40°C indicate that the n-paraffins encapsulated in high density polyethylene will retain at least 90% of the n-paraffins. The encapsulated n-paraffins have been incorporated into cellulose insulation and exposed to outside conditions for over a month. Results include a more than 10% reduction in heat flux through insulation with our PCM and a 1°C reduction in peak temperature compared with plain insulation. An economic analysis will be presented as well.

Sustainability through Durable Building Materials and Assemblies

Dr. Theresa Weston
DuPont Building Innovations

An opportunity for chemical engineers in transforming the sustainable or green building industry is in developing understanding of building durability. Building durability is a key sustainability attribute. A building cannot be truly energy efficient or sustainable unless it lasts. One of the most important components of building durability is the prevention of moisture damage. Chemical engineering knowledge of combined heat and mass transfer can provide key insights in to moisture damage risk analysis, and can therefore aid in the design on energy efficient and durable buildings.

Sustainable Building Codes & Standards

Despite the overall slow-down of the construction industry, especially the home building industry in North America, one growth area has been green and energy efficient construction. This trend has coincided with a large number of changes in the standards, codes and regulations governing the building construction. More specifically, the industry has introduced new energy efficiency and sustainability codes and standards. While these codes in many ways are an extension of existing building codes, the combination of more stringent, and in some cases new, requirements, lead to the use of new building materials, assemblies and systems. Although a significant amount of research has been conducted moisture durability of traditional wall systems, this research needs to be extended to apply to these new energy efficient and sustainable building systems

Modeling Moisture Damage Risk

Modeling combined heat and mass (water) transport or hygrothermal modeling is a useful tool, which is becoming more prevalent in the construction industry. This type of modeling can highlight the risk of moisture accumulation in building systems. An example (shown below) is the higher risk of accumulation of water from building assemblies when additional layers of low vapor permeability materials such as exterior insulation are added to a wall system.

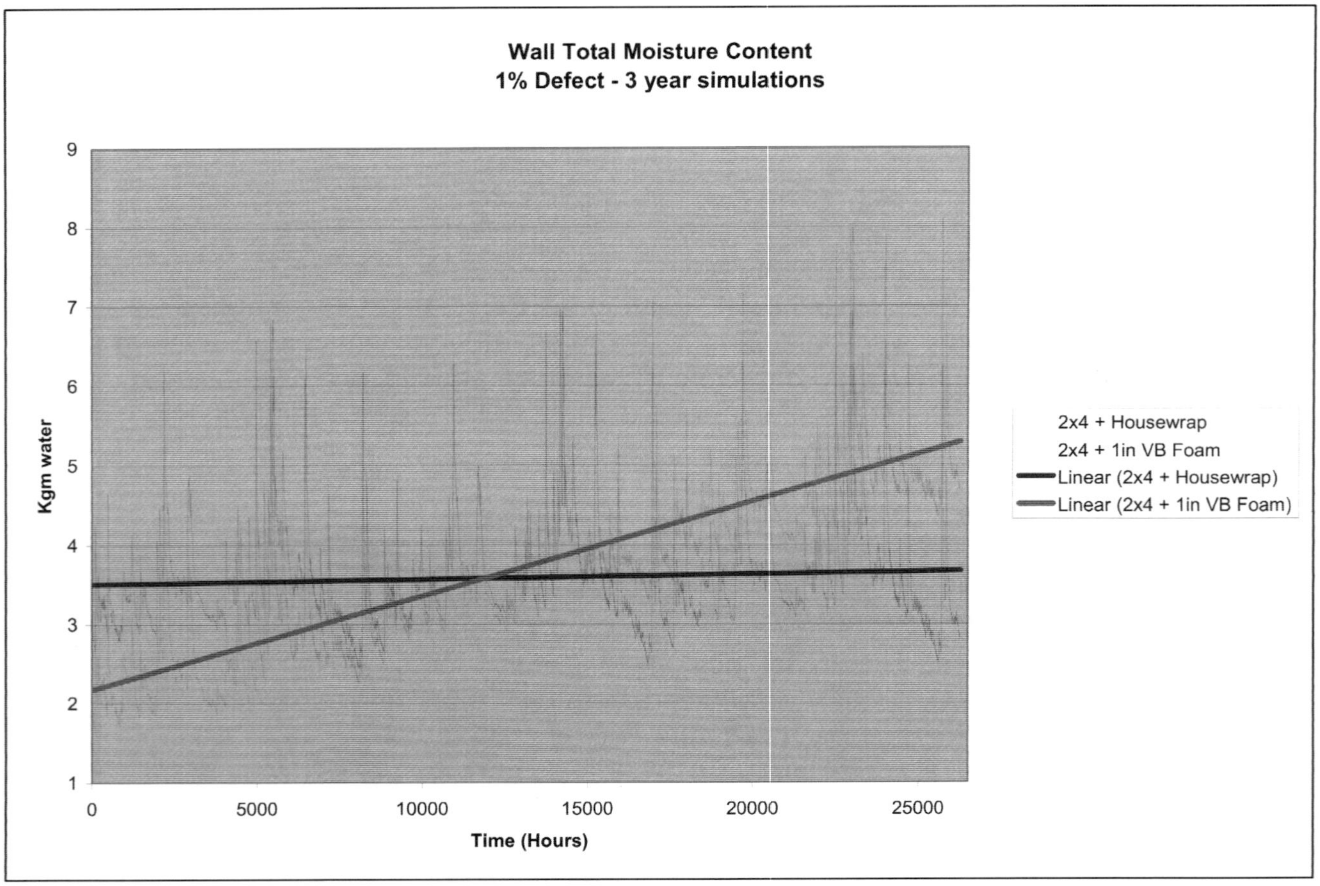

This presentation will provide information on the status of moisture modeling in the construction industry including:

- Standardization of protocols for the use of moisture risk analysis by the development of ASHRAE Standard 160 – *Criteria for Moisture-Control Design Analysis in Buildings*

- Sponsorship of hygrothermal modeling software development by the US Department of Energy, and

- Development of new methodologies to extend the use of hygrothermal simulation to include the management of drying of incidental water leaks.

Examples of specific materials used to help provide durable, energy efficiency to buildings will be discussed.

AUTHOR INDEX

AUTHOR INDEX

American Institute of Chemical Engineers
3 Park Avenue
New York, New York 10016-5991

proceedings.com